COPERNICUS AND HIS SUCCESSORS

COPERNICUS
AND HIS SUCCESSORS

EDWARD ROSEN

THE HAMBLEDON PRESS
LONDON AND RIO GRANDE

Published by The Hambledon Press 1995

102 Gloucester Avenue, London NW1 8HX (U.K.)

P.O. Box 162, Rio Grande, Ohio 45674 (U.S.A.)

ISBN 1 85285 071 X

A description of this book is available from
the British Library and from the Library of Congress

Printed on acid-free paper and bound in
Great Britain by Cambridge University Press

Contents

Acknowledgements

The articles reprinted appeared in the following places and are reprinted here by the kind permission of the original publishers.

1   *Bulletin of the American Society of Papyrologists* (1978), pp. 85–93.

2   *Isis*, 53 (1962), pp. 504–8.

3   *Didascaliae: Studies in Honour of Anselm M. Albareda*, edited by Sesto Prete (Bernard M. Rosenthal, New York, 1961), pp. 369–79.

4   *Manuscripta*, 20 (1976), pp. 163–74.

5   *Centaurus*, 7 (1961), pp. 152–56.

6   *Archiv für Reformationsgeschichte*, 62 (1971), pp. 90–98.

7   *Sixteenth-Century Journal*, 12 (1981), pp. 13–16.

8   *Journal of the History of Ideas*, 32 (1971), pp. 281–88.

9   *Dialectics and Humanism*, 1 (1987), pp. 257–65.

10 *Organon*, 2 (1965), pp. 61–78.

11 *Centaurus*, 20 (1976), pp. 44–49.

12 *Sudhoffs Archiv*, 61 (1977), pp. 144–55.

13 *Archives Internationales d'Histoire des Sciences*, 25 (1975), pp. 82–91.

14 *Accademia Nazionale dei Lincei*, 216 (1975), pp. 27–37.

15 *Physis*, 23 (1981), pp. 449–57.

16 *Journal of the History of Ideas*, 36 (1975), pp. 531–42.

17 *Journal of the History of Ideas*, 21 (1960), pp. 431–41.

18 *Centerpoint* (Fall, 1976), pp. 46–55.

19 *Saggi su Galileo Galilei* (Barbera, Florence, 1967), pp. 1–12.

20 *Isis*, 32 (1958), pp. 319–30.

21 *Papers of the Bibliographical Society*, 75 (1981), pp. 401–12.

22 *Vistas in Astronomy*, 18 (1975), pp. 225–31.

To
Adam Kanarek
For his invaluable help
throughout the years

Introduction

The late Dr Edward Rosen, Distinguished Professor Emeritus of the History of Science at the Graduate School of the City University of New York, was for nearly half a century not only one of the foremost Copernican scholars, but he also achieved great distinction in dealing with three of Copernicus's most famous successors, that is Tycho Brahe, Johannes Kepler and Galilei Galileo. For his lifelong painstaking researches on Copernicus, Rosen was awarded a Gold Medal of Merit from the Polish People's Republic (that country's highest award bestowed on a private citizen of a foreign country), and Medal of the Copernicus Society of America, while the History of Science Society presented Rosen with the likewise very prestigious Pfizer Medal for his book *Kepler's Somnium*.

A prolific writer of several books and translations, as well as hundreds of articles published in a variety of scholarly journals, Rosen was beseeched by scholars from around the globe for the offprints of his essays, as some of them were not easy to obtain. These requests did not cease until the present time, but since the supply is completely exhausted, I am grateful to Martin Sheppard, the publisher of the Hambledon Press, for collecting twenty-two of Rosen's most sought after essays and making them available in one volume.

Although the book is aptly titled *Copernicus and his Successors*, several of the essays deal also with Copernicus's predecessors, while others give – in my opinion – definitive solutions to problems still heatedly disputed by copernicologists (see for example 'Copernicus' Axioms', and 'Copernicus' Spheres and Epicycles').

The essay titled 'Copernicus' Alleged Priesthood' disposes once for all with the notion that Copernicus was an ordained priest. And 'Copernicus Was Not a "Happy Notary"' puts an end to the claim, propagated recently by Varmia's Bishop Jan Obłąk, that Copernicus was a notary public.

To reiterate the statement made by Johannes Petreius of Nuremberg, the 1543 publisher of Copernicus's major treatise *De revolutionibus orbium coelestium*: 'eme, lege, et fruere' (buy, read, and enjoy [this work]).

Erna Hilfstein

1

Aristarchus of Samos and Copernicus

A famous passage about Aristarchus of Samos is found in Archimedes' *Sand-Reckoner*:

> According to Aristarchus' hypotheses, the fixed stars and the sun remain stationary. On the other hand, the earth revolves around the sun along the circumference of a circle lying amid the course [of the planets]. The sphere of the fixed stars has the same center as the sun [and is virtually infinite].

These propositions, formulated with magisterial succinctness by a great mathematician, contain the essence of the heliostatic, heliocentric, and geokinetic system later propounded by the founder of modern astronomy, Nicholas Copernicus (1473–1543). Indeed, the striking similarity between the two sets of concepts, ancient and modern, led to the labeling of Aristarchus as "The Ancient Copernicus."[1]

What did the modern Copernicus know about the ancient Copernicus? In particular, was the modern Copernicus familiar with the discussion of Aristarchus in Archimedes' *Sand-Reckoner*?

An all but completely affirmative answer to this question was given thirty-six years ago by two authors, who asserted "the almost certain acquaintance of Copernicus with the *Sand-Reckoner*."[2] These two "authors have made it very plausible that Copernicus knew the *Sand-Reckoner* and could thus have been influenced by Aristarchus' ideas," according to a well-known historian of ancient astronomy.[3] These authors "argued the almost certain acquaintance of Copernicus with

---

1 Thomas Heath, *Aristarchus of Samos, the Ancient Copernicus* (New York 1959; reprint of Oxford 1913 edition).

2 Rudolf von Erhardt and Erika von Erhardt-Siebold, "Archimedes' Sand-Reckoner, Aristarchos and Copernicus," *Isis* 33 (1941–1942) 578.

3 Otto Neugebauer, "Archimedes and Aristarchus," *Isis* 34 (1942–1943) 4.

Archimedes' *Arenarius,* the work that contains the most authoritative and best account of Aristarchus' theory," according to a recent study.[4]

How could Copernicus become acquainted with Archimedes' *Sand-Reckoner?* Its Greek text was first published in March 1544, nearly a year after Copernicus died on 24 May 1543. Consequently, he never saw the *Sand-Reckoner* printed in Greek. This first edition of Archimedes' works in Greek, with the *Sand-Reckoner* at pages 120–127, was accompanied by a separately paginated translation of these works into Latin, with the first printed Latin translation of the *Sand-Reckoner* at pages 155–163. By the same token, therefore, Copernicus never saw a Latin version of the *Sand-Reckoner* in print.

Did he see a manuscript copy of the *Sand-Reckoner,* either in Greek or in Latin or in both classical languages? The foremost astronomer of the fifteenth century, Johannes Regiomontanus (1436–1476), copied with his own hand the Greek text, as well as a Latin version, of the *Sand-Reckoner* while he was in Italy. From that country Regiomontanus took his Archimedes manuscript with him to Nuremberg, where it is still preserved in the municipal library.[5] Did Copernicus see this Regiomontanus manuscript? There is no indication that Copernicus ever set foot in Nuremberg. If he had, and if he knew about the Aristarchus passage in the *Sand-Reckoner,* and if he had heard of the Regiomontanus manuscript, he still would have had no access to it. For after the death of Regiomontanus, his papers passed to his pupil Bernard Walther (about 1430–1504), "a melancholy and taciturn man, who withheld them from use and refused to let anyone so much as inspect them. Even after Walther's death in 1504 the books and instruments . . . remained for some time in the hands of the executors."[6] Surely Copernicus never saw the *Sand-Reckoner*'s Aristarchus passage in Regiomontanus' Archimedes manuscript in Nuremberg.

But what about Italy, where Copernicus was a student from 1496 to 1503? The aforementioned two authors, who asserted "the almost certain acquaintance of Copernicus with the *Sand-Reckoner,*" also stated that Copernicus "at least knew of Giorgio Valla, who died in 1499."[7] Valla died on 23 January; the year was 1499 according to the Venetian calendar

---

4   William Harris Stahl, *Martianus Capella and the Seven Liberal Arts* (New York and London 1971) 176.

5   Nuremberg, Stadtbibliothek, Cen V, 15, with the *Sand-Reckoner* at folios 161–172; see Ernst Zinner, *Leben und Wirken des Johannes Müller von Königsberg genannt Regiomontanus,* 2nd ed. (Osnabrück 1968) 322–323.

6   Hans Rupprich, "Willibald Pirckheimer," in *Pre-Reformation Germany,* ed. Gerald Strauss (London 1972), p. 412.

7   P. 600 of the work cited in note 2, above.

then in use, but 1500 according to the present calendar.[8] Not only did Copernicus know of Valla, he also made extensive use of Valla's posthumous *De expetendis et fugiendis rebus,* which was published in Venice in December 1501. Although Valla included his own translation of some of Archimedes' writings in this massive posthumous encyclopedia, the *Sand-Reckoner* was completely omitted from its forty-nine Books.[9] Therefore, neither in Italy nor in Nuremberg, neither from Regiomontanus nor from Valla, did Copernicus learn about the *Sand-Reckoner*'s invaluable discussion of Aristarchus.

The external evidence that Copernicus did not know about Aristarchus' heliocentrism from Archimedes' *Sand-Reckoner* is confirmed by the internal evidence provided by Copernicus' own writings. He never claimed that he was the first geokineticist. In fact, he searched for and named as many of his predecessors as he could trace. In the Dedication of his *Revolutions (De revolutionibus orbium coelestium* [Nuremberg 1543]) he announced:

> At first I found in Cicero that Hicetas supposed the earth to move. Later I also discovered in Plutarch [Copernicus refers to Pseudo-Plutarch, *The Opinions of the Philosophers*] that others were of this opinion. I have decided to set his words down here, so that they may be available to everybody: "Some think that the earth remains at rest. But Philolaus the Pythagorean believes that, like the sun and moon, it revolves around the fire in an oblique circle. Heraclides of Pontus and Ecphantus the Pythagorean make the earth move, not in a progressive motion, but like a wheel in a rotation from west to east about its own center."

Moreover, in Book I, Chapter 5, of the *Revolutions* Copernicus remarked:

> Heraclides and Ecphantus, the Pythagoreans, as well as Hicetas of Syracuse... rotated the earth in the middle of the universe.... That the earth rotates, that it also travels with several motions, and that it is one of the heavenly bodies, are said to have been the opinions of Philolaus the Pythagorean.

From these two retrospective passages of the *Revolutions* dealing with the ancient geokineticists, Aristarchus is conspicuously absent. For this absence, two explanations are available:

(1) Copernicus had no satisfactory evidence permitting him to group Aristarchus with the other four geokineticists; or

---

8   J. L. Heiberg, *Beiträge zur Geschichte Georg Vallas und seiner Bibliothek* (Leipzig 1896) 41 (= XVI. *Beiheft zum Zentralblatt für Bibliothekswesen*).

9   For Valla's Greek sources, see J. L. Heiberg, "Philologische Studien zu griechischen Mathematikern, III, Die Handschriften Georg Vallas von griechischen Mathematikern," *Jahrbücher für classische Philologie,* 1881, Supplementband P12, 375–402, especially 380–386.

(2) Copernicus possessed such evidence, and deliberately suppressed it.

The second alternative is based on a deleted passage in the autograph manuscript of Copernicus' *Revolutions*. This manuscript is now preserved in the Jagellonian Library of the University of Cracow, where Copernicus was a student in his late teens and early twenties. A photofacsimile of his autograph manuscript was recently published.[10] It plainly shows the deleted passage (folio 11 verso), which said, in part:

> Philolaus was aware that the earth could move. Aristarchus of Samos was also of the same opinion, as some people say.

These deleted lines make quite clear Copernicus' unfamiliarity with the Aristarchus passage in the *Sand-Reckoner*. Had Copernicus realized that he could cite Archimedes as his authority for Aristarchus as a fifth geokineticist, he surely would not have submerged that renowned mathematician in an anonymous and undistinguished group of "some people" (*nonnulli*). Nor would Copernicus have confined his remark about Aristarchus to the earth's mobility, had he learned about the additional features of Aristarchus' astronomy from Archimedes' report: the sun's immobility and centrality, the stars' immobility and enormous remoteness. Had Copernicus known that he could align the great Archimedes on his side, that he could add the distinctively Aristarchan insights to those proclaimed by the other four geokineticists, he would have leaped with joy. For he was painfully aware that, with the theologians and Aristotelian philosophers certain to denounce him, he needed all the support he could muster.

Copernicus' explicit citations of Cicero and [Pseudo-] Plutarch point to the same conclusion as the history of the *Sand-Reckoner* in manuscript and in print: Copernicus was not familiar with that work by Archimedes. The Syracusan's numerous treatises on the mathematical sciences were virtually a closed book for Copernicus, who mentioned Archimedes in only three connections:

(1) Archimedes made the year $365\frac{1}{4}$ days long (*Revolutions, Book 3, Chapter 13*);
(2) Archimedes used a square in determining the area of a circle (Book 3, Chapter 13, a deleted passage); and
(3) Archimedes set limits to the value of π (Book 4, Chapter 32).

The length of the year, the area of a circle, and the value of π were not discussed in the *Sand-Reckoner*. Nothing in Copernicus' writings suggests that he knew Archimedes' *Sand-Reckoner*, or that he ever heard of it.

10   Nicholas Copernicus, *Complete Works*, I (London/Warsaw 1972).

Without access to Archimedes' *Sand-Reckoner,* how did Copernicus learn that, according to Aristarchus, the "earth could move"? A copy of Aristarchus' own presentation of his geokineticism found its way into the hands of Archimedes, who was some twenty-five years his junior. With reference to this copy, Archimedes said in the *Sand-Reckoner:* "Descriptions of some hypotheses were made public by Aristarchus." Unfortunately, those descriptions did not survive. Copernicus never saw them, any more than he ever saw Archimedes' account of them in the *Sand-Reckoner.*

But Copernicus did see the statement that "the earth moves" attributed to Aristarchus in Pseudo-Plutarch's *Opinions of the Philosophers,* Book 2, Chapter 24. From Book 3, Chapter 13, of that work he quoted in the original Greek the passage in his Dedication of the *Revolutions* that was translated above. Unquestionably, therefore, Copernicus had a copy of the Greek text of Pseudo-Plutarch's *Opinions.*

Yet Copernicus did not quote from its Book 2, Chapter 24, as he did from its Book 3, Chapter 13. Why not? According to Book 2, Chapter 24, "Aristarchus makes the sun stationary among the fixed stars." Today we are not shocked by a stationary sun among the stars. But for the ancients and Copernicus, the sun was much closer to the earth than the stars were. As though an intrastellar sun were not strange enough, *Opinions,* Book 2, Chapter 24, also attributed to Aristarchus the opinion that "the earth moves around the solar circle." The expression "solar circle" was left unexplained in the *Opinions.* Perhaps it referred to the sun's annual circular orbit in the pre-Aristarchan astronomy, which held the central earth stationary. If so, Aristarchus' mobile "earth moves around the solar circle" of the pre-Aristarchans. Then, the earth's orbit would be somewhere among the stars. This weird cosmos did not appeal to Copernicus, a practical astronomer with both his feet planted firmly on solid ground. In Copernicus' judgment, Aristarchus forfeited his membership on the team of respectable ancient geokineticists by locating his moving earth among the stars.

Since Copernicus quoted the original Greek of *Opinions,* Book 3, Chapter 13, he had access to a copy of *Plutarchi opuscula LXXXXII* (Venice 1509). He likewise used Valla's *De expetendis et fugiendis rebus,* where Books 20–21 were Valla's own translation of the *Opinions.* In his characteristic way, however, Valla gave no indication that his Books 20–21 were his own rendering of a Greek text. From Valla, Copernicus took numerical values and expressions which he repeated in his own *Revolutions,* Book 1, Chapter 10. In Book 3, Chapter 2, Copernicus made a mistake previously made only by Valla. Copernicus' use of Valla is therefore indisputable. But Valla's Book 21, Chapter 24, was even worse than the Greek text of the *Opinions.* For, Valla had Aristarchus "locate the sun beyond the fixed stars." In general, the intellectual level of Valla's vast encyclopedia was so low that Copernicus never mentioned Valla's name. But Copernicus had great respect for Plutarch, whom he (mistakenly) regarded as the author of the *Opinions*. Nevertheless, Book 2, Chapter 24, of the *Opinions* was so obviously garbled that Copernicus would not attribute that mess to Plutarch. Hence he resorted to the conveniently vague formula "some people" when he mentioned Aristarchus' geokineticism in the passage he later deleted.

Why did he delete this passage? Was Copernicus trying to hide any indebtedness to Aristarchus? If so, why did he allow Book 1, Chapter 5, to be printed with its attribution to Philolaus of the opinion that the earth is a heavenly body endowed with several motions, including rotation? Was Philolaus respectable, and Aristarchus not?

Copernicus quoted from *Plutarchi opuscula LXXXXII*. This volume contained, besides the spurious *Opinions*, Plutarch's authentic dialogue on *The Face in the Moon*. Its interlocutors were reminded that a famous philosopher believed that it was the duty of the Greeks "to indict Aristarchus of Samos for impiety in imputing motion to the Hearth of the Universe." These interlocutors needed no reminder that in Plato's *Phaedrus* (247A) the expression for the immobility of the earth was: "in the house of the gods, the Hearth stands still." In Plutarch's *Face in the Moon*, the indictment continued:

> According to Aristarchus' hypotheses, the heaven stands still whereas the earth revolves along an oblique circle. At the same time the earth rotates around its own axis.

The earth's rotation was not explicitly mentioned as an integral feature of Aristarchus' astronomy in Archimedes' *Sand-Reckoner*, which rather focused attention on the earth's revolution around the sun. The only clear statement that Aristarchus' earth also rotated occurs in Plutarch's *Face in the Moon*. Had Copernicus known this passage, he would have specified how Aristarchus' earth moved. Instead, he said nothing more than that it moved. Hence, Copernicus did not know Plutarch's *Face in the Moon*.

Its passage about Aristarchus appears on page 932 in *Plutarchi opuscula LXXXXII*. From page 328 in that volume Copernicus took the Greek quotation in the Dedication of his *Revolutions*. For him, the doxographical *Opinions* had great importance. The rest of that bulky volume was not used by Copernicus, a busy canon with very little spare time for reading. He would have advanced his cause considerably had he not overlooked the valuable reference to Aristarchus in Plutarch's *Face in the Moon*. Copernicus' only reference to Aristarchus' geokineticism was deleted with the rest of the deleted passage. But the purpose of the deletion was not to get rid of the name of Aristarchus, who was mentioned elsewhere in the *Revolutions*.

The deleted passage consists of about eight hundred words, in exactly ten of which Aristarchus is mentioned or in about 1¼% of the entire deletion. Had Copernicus wanted to get rid of a potentially dangerous association with Aristarchus, all he had to strike out was those ten words. Their presence in the deleted passage is of such minor significance that removing them would not have impaired the rest of the passage in any way whatsoever. The bulk of the passage consists of Copernicus' translation into Latin of an ancient Greek letter. Its text had already been printed. Then Copernicus acquired a copy of a book containing an expert Latin translation of the letter by a highly reputable authority. What would be the sense of publishing a second translation of the letter? Instead of doing so, in the part of the *Revolutions* that he wrote last, Copernicus briefly summarized the contents of the letter while suppressing his translation of it.

With this deletion, Aristarchus disappeared from the *Revolutions* as a geokineticist. But he remained as having "found the same obliquity of the ecliptic and equator as Ptolemy, 23°51′20′″" (Book III, Chapter 2). "According to Ptolemy, the obliquity of 23°51′20″ did not change at all in the four hundred years from Aristarchus of Samos to his own time" (Book III, Chapter 6). "The natural year, which is also called the tropical year, is determined by Aristarchus of Samos as containing 365¼ days" (Book III, Chapter 13). Neither of these two values, 23°51′20″ and 365¼ days, can be found in Aristarchus' extant writings. Nor were they ever attributed to Aristarchus by anybody before Copernicus. Why did he make these two mistakes?

Copernicus says: "according to Ptolemy." What access to Ptolemy did Copernicus have? The Greek text of Ptolemy's *Syntaxis* was first printed in 1538. A copy reached Copernicus too late to exert any significant effect on the composition of the *Revolutions*. Lacking the Greek text, whether in print or in manuscript, Copernicus perforce had to rely on a Latin translation. In the main he used the translation published in Venice in 1515. This was not a recent rendering based on Ptolemy's Greek text and made by a scholar who knew both Greek and astronomy. Instead, it was a medieval translation based, not on Ptolemy's Greek text, but on a previous translation from Greek into Arabic. In passing in this way from Greek into Arabic, and then from Arabic into Latin, Ptolemy suffered grievously.

In particular, names of Greek astronomers mentioned by Ptolemy became distorted. Thus in Book VII, Chapter 1, the 1515 edition referred to observations by "Arsatilis" and Timocharis. What Greek astronomer lay behind "Arsatilis"? Copernicus guessed "Aristarchus." In his personal copy of the 1515 edition, which is now in the library of the University of Uppsala in Sweden, Copernicus underlined "Arsatilis" on folio 75 verso (Book VII, Chapter 3) and replaced it by Aristarchus in the margin.[11] In a letter dated 3 June 1524, Copernicus referred to "Aristarchus and Timocharis," and he also said that Ptolemy "joined Aristarchus to Timocharis of Alexandria as his contemporary."[12]

Later Copernicus discovered the correct equation, Arsatilis = Aristyllus. On folio 73 recto (Book VII, Chapter 1) of his copy of the 1515 Ptolemy, Copernicus made the proper substitution of Aristyllus for Arsatilis.[13] By the same token, in his autograph manuscript of the *Revolutions* (folio 78 verso, line 2), where Copernicus had originally written "Aristarchus," he replaced that misnomer by the appropriate name, Aristyllus.

On the other hand, Copernicus left "Aristarchus" unchanged in two other related passages (folio 73 recto, lines 20, 28, and folio 79 recto, line 16). When Copernicus had to make a change somewhere in his autograph manuscript, he did not always systematically hunt for the other passages where corresponding changes were then required. As a result of this erratic procedure, his autograph is riddled with many internal inconsistencies.

Copernicus discovered the blunder in his equation Arsatilis = Aristarchus after 3 June 1524. He may have seen the light after getting hold of a copy of the 1528 Latin translation of the *Syntaxis*. This was made directly from the Greek, without any Arabic detour. Some of Copernicus' expressions (*Revolutions*, Book I, Chapter 7) unmistakably echo the 1528 translation.

This helped him somewhat, but did not clear up all the difficulties. For instance, the 23°51'20" obliquity was really ascribed by Ptolemy to Eratosthenes, not to Aristarchus. In like manner, the 365¼ day length of the year was attributed to Callippus, which became "felis" in the 1515 edition. The same length of the year was misattributed by Copernicus to Archimedes ("Arsamides" in the 1515 edition). The latter might have made an error of ¼ day, Hipparchus ("Abrachis") surmised. Somehow, a possible error of ¼ day was misread as 365¼ days as the length of the year.

Poor Copernicus, his imperfect sources led him to confuse Aristarchus with Eratosthenes and Callippus. He never quite achieved a clear separation of Aristarchus from Aristyllus. At best, his picture of Aristarchus of Samos was sadly out of focus. That is why his greatest heliocen-

11   Ludwik Antoni Birkenmajer, *Mikołaj Kopernik* (Cracow 1900) 256, no. 64.
12   Edward Rosen, *Three Copernican Treatises,* 3rd ed. (New York 1971) 101, 103.
13   Birkenmajer 256, no. 59.

trist predecessor, Aristarchus, the first astronomer to make the earth revolve as a planet around a stationary sun, failed to appear alongside others of like mind in the printed version of the *Revolutions*.

2

Was Copernicus a Pythagorean?

> "Let no one suppose that I have gratuitously asserted, with the Pythagoreans, the motion of the earth."
> Copernicus, *Commentariolus*, Assumptions.

In a recent article Professor Thomas W. Africa maintained that Copernicus saw his thesis "as the foundation of a Pythagorean Restoration." [1] In support of this contention Professor Africa offered no evidence whatever. Instead, he cited Kepler and Galileo. But both of these great Copernicans were born decades after Copernicus died. Galileo made a number of egregious errors regarding Copernicus. [2] Although Kepler did have access to some of Copernicus' correspondence which is no longer available to us, his thinking was far from a carbon copy of Copernicus'. Therefore no opinion of Kepler may

legitimately be imputed retroactively to Copernicus.

Having reminded ourselves that Copernicus' mentality was quite different from Galileo's and Kepler's, let us now see whether the statements quoted from them by Professor Africa mean that Copernicus intended to restore Pythagoreanism. Kepler referred to "Pythagoras, grandfather of all Copernicans." This little joke proclaims a remote ancestral intellectual relationship between Pythagoras and Copernicus; it does not ascribe to Copernicus any intention to restore Pythagoreanism. By the same token, Galileo labeled the belief in "the earth's mobility a Pythagorean opinion." But in using this entirely correct label, Galileo did not attribute to Copernicus any intention to restore Pythagoreanism. According to Professor Africa (p. 404), "Copernicus insisted that his system was . . . the revival of . . . the lost doctrine of Pythagoras." If a doctrine is lost, how can it be revived? When and where did Copernicus insist that he was reviving a doctrine that could not be

---

[1] "Copernicus' Relation to Aristarchus and Pythagoras," *Isis*, 1961, *52*: 403.

[2] Edward Rosen, "Galileo's Misstatements about Copernicus," *Isis*, 1958, *49*: 319-330; reprinted as Massachusetts Institute of Technology, Publications in the Humanities, no. 32 (1958) ; below, 193–204.

revived? No citation to this effect appears in the fifty-seven footnotes accompanying Professor Africa's seven-page article.

Professor Africa asserted (p. 406) that "in the manuscript of *De revolutionibus*, Copernicus frankly professed Pythagoreanism as he conceived it, but deleted the passage." Actually, in the deleted passage Copernicus did not profess Pythagoreanism. On the contrary, he discussed it without professing it. Nor is it true that here "Copernicus associated Aristarchus and Philolaus with a view of the earth's motion other than the traditional Pythagorean scheme" (p. 407). Copernicus' language provides no basis for distinguishing between the older and later Pythagorean schemes. Professor Africa himself declared (p. 408) that Copernicus "felt that he was restoring the Pythagorean astronomy in its purity." Professor Africa then proceeded to mistranslate a statement by Copernicus' first disciple, whose reference to "the Pythagoreans" was altered by Professor Africa to "Pythagoras." Professor Africa printed this incorrect version in such a way as to imply that the error occurs in Edward Rosen's *Three Copernican Treatises*, which Professor Africa cited according to the first edition, instead of the second and enlarged edition.[3]

Having translated the deleted passage (after a fashion), Professor Africa maintained (pp. 407-408) that it "revealed Copernicus' conviction that . . . Plato shared the scientific views of Philolaus." But the deleted passage says no such thing. It does say that "according to Plato, there were only a few philosophers who at that time knew how to explain the motions of the planets." Just before this strange distortion of Copernicus' simple Latin, Professor Africa charged (p. 407) that "Copernicus drew an unwarranted assumption from Plato's notion that 'true' astronomy rested on intuitive reason." Did Copernicus ever draw an assumption, warranted or unwarranted, from a notion? If he ever did,

how are we to tell whether this particular assumption was unwarranted? Professor Africa somehow neglected to tell us what the unwarranted assumption was.

In addition to gratuitously accusing one of the greatest astronomers in mankind's history of not thinking straight, Professor Africa undertook to sit in judgment on Copernicus' personality as a scientist. Thus Professor Africa has decided (pp. 403-404) that "Copernicus lacked . . . concern for data and method." He has aimed this grave indictment at the man whose first disciple, having worked with him nearly three years, reported that

my teacher always has before his eyes the observations of all ages together with his own, assembled in order as in catalogues; then when some conclusion must be drawn or contribution made to the science and its principles, he proceeds from the earliest observations to his own, seeking the mutual relationship which harmonizes them all; the results thus obtained by correct inference under the guidance of Urania he then compares with the hypotheses of Ptolemy and the ancients; and having made a most careful examination of these hypotheses, he finds that astronomical proof requires their rejection; he assumes new hypotheses, not indeed without divine inspiration and the favor of the gods; by applying mathematics, he geometrically establishes the conclusions which can be drawn from them by correct inference; he then harmonizes the ancient observations and his own with the hypotheses which he has adopted; and after performing all these operations he finally writes down the laws of astronomy.[4]

The foregoing passage occurs in Rheticus' *First Report* (*Narratio prima*). To Professor Africa (p. 403), "the tone of Rheticus' *Narratio prima* suggests that Copernicus' attitude toward other scientists was generally . . . contempt." Let us look at the *First Report*'s attitude toward other scientists: Domenico Maria Novara is "very learned"; Copernicus "knows that Ptolemy, working with the utmost care, . . . accurately investigated the motions of the sun and the moon and established them correctly, so far as he could"; "Surely students hereafter will see the value of Ptolemy and the other ancient writers, so that they will recall

[3] New York: Dover; London: Constable, 1959.

[4] *Three Copernican Treatises*, p. 163.

these men who have been until now excluded from the schools, and restore them, like returned exiles, to their ancient place of honor "; " How modestly and wisely Aristotle speaks on the subject of the celestial motions can be seen everywhere in his works "; " Pliny set down an excellent and accurate statement "; " Plato and the Pythagoreans, the greatest mathematicians of that divine age "; " that wise statement of Socrates." [5] Where in the *First Report* is the tone which suggested to Professor Africa " that Copernicus' attitude toward other scientists was generally . . . contempt "?

Professor Africa (p. 406) proposed " vanity " as a possible motive for one of Copernicus' most significant actions. But Rheticus tells us that Copernicus " promised that he would draw up astronomical tables with new rules and that if his work had any value he would not keep it from the world." [6] Does that sound like " vanity "?

Professor Africa (p. 403) referred to " Copernicus' imperious *Letter against Werner*." Let us see how imperious it is: " Faultfinding is of little use and scant profit. . . . Hence I fear that I may arouse anger if I reprove another while I myself produce nothing better. . . . It is one thing to snap at a man and attack him, but another thing to set him right and redirect him when he strays." [7] Is this letter " imperious," that is, arrogant or haughty?

According to Professor Africa (p. 405), in the dedication of the *Revolutions* " Copernicus was not entirely frank, for his obligation to the ancients included more than simple suggestions of the earth's mobility." But in the dedication Copernicus had occasion to describe only this part of his obligation to the ancients. Elsewhere he described other parts. For instance, in the *Letter against Werner*, which Professor Africa termed " imperious," Copernicus wrote with regard to the ancient mathematicians:

we must follow in their footsteps and hold fast to their observations, bequeathed to us

like an inheritance. . . . It is well known that they observed all these phenomena with great care and expert skill, and bequeathed to us many famous and praiseworthy discoveries.[8]

On the one hand, Professor Africa held that " Copernicus frankly professed Pythagoreanism " where he did not profess it, whether frankly or not. On the other hand, Professor Africa concluded that " Copernicus was not entirely frank " because he chose to put a partial statement in its proper place. In order to be able to determine whether another human being was or was not entirely frank, the self-appointed judge should first try to acquire some understanding of his victim's character and thought.

It might also be helpful if he learned about his victim's career and associates. Professor Africa referred (p. 405) to a teacher of philosophy as Copernicus' " colleague at Ferrara," where Copernicus never taught. And Professor Africa (p. 405) spoke of " Cordus Urceo," when he meant Antonio Urceo, who playfully adopted the nickname " Codro " (the pauper).[9]

A would-be judge might likewise be expected to inform himself about his victim's reputation. According to Professor Africa (pp. 403-404), " Copernicus lacked the concern for data and method which partly prevented Tycho Brahe from accepting the Copernican system." This ungrammatical verdict seems intended to convey the idea that Brahe had a low opinion of Copernicus' " concern for data and method." Since Professor Africa again failed to provide us with a citation, let us listen to Brahe's famous oration:

In our time Nicholas Copernicus may not undeservedly be called a second Ptolemy. Through observations made by himself he discovered certain gaps in Ptolemy, and he concluded that the hypotheses established by Ptolemy admit something unsuitable in violation of the axioms of mathematics. Moreover, he found the Alfonsine computations in disagreement with the motions of the heavens. Therefore, with wonderful intellectual acumen he

---

5 *Op. cit.*, pp. 111, 126, 132, 141, 144, 147, 168.
6 *Op. cit.*, p. 192.
7 *Op. cit.*, p. 93.

8 *Op. cit.*, pp. 99-100.
9 Edward Rosen, " Copernicus' Quotation from Sophocles," in *Didascaliae, Studies in Honor of Anselm M. Albareda*, ed. Sesto Prete (New York, 1961), p. 372; below, p. 20 .

established different hypotheses. He restored the science of the heavenly motions in such a way that nobody before him had a more accurate knowledge of the movements of the heavenly bodies.[10]

Brahe's reasons for rejecting Copernicanism had nothing to do with " data and method." He found the new system in conflict with " not only the principles of physics but also the authority of Holy Scripture." [11] Brahe's biographer and editor of his collected works tells us that Tycho never mentioned Copernicus' name " without some expression of admiration." [12]

The aforementioned deleted passage was characterized by Professor Africa as being " incongruous in an astronomical tract " (p. 407). Why did Copernicus write the passage in the first place, and why did he later delete it? He had to explain why he kept his treatise hidden so long. For this purpose he recalled the Pythagoreans' practice of communicating their knowledge only to friendly persons. When he was finally persuaded to release his manuscript to a printer, he drafted the dedication to the pope. In the dedication he related that he had debated with himself for a long time whether to publish his treatise or " follow the example of the Pythagoreans and certain others." His discussion of their example made the earlier passage redundant, and consequently he deleted it. But, despite Professor Africa, it was never incongruous. On the contrary, either it or some equivalent was indispensable to account for Copernicus' long delay in publishing his book.

According to Professor Africa (p. 403), Copernicus' " hesitation in publishing the *Revolutions* has often been ascribed to fear of ecclesiastical opposition, but the theme of sacerdotal oppression is difficult to maintain in view of demonstrated papal favor. In 1536 Pope Paul III extended encouragement to Copernicus through Cardinal von Schönberg." How was the papal favor

demonstrated? What is the basis of Professor Africa's pronouncement that " in 1536 Pope Paul III extended encouragement to Copernicus through Cardinal von Schönberg "? Certainly the cardinal's letter of 1536 to Copernicus said nothing about papal encouragement or favor. Nor did Copernicus, in dedicating the *Revolutions* to Pope Paul III, say anything about any papal encouragement or favor in 1536 or any other year. Copernicus gave the first place in the *Revolutions* to the cardinal's letter. Next came the dedication to the pope. Here a bishop was singled out as having frequently exhorted Copernicus to publish. Amid all this impressive parade of ecclesiastical dignitaries, Copernicus made no mention of any papal encouragement or favor. Had he ever received any, would he not have displayed it as conspicuously as possible? Has the argument from silence ever been more eloquent?

We have now seen that there is absolutely no historical foundation for Professor Africa's allegation about Copernicus' receiving papal favor or encouragement. Let us next consider how well Professor Africa understood why Copernicus hesitated to publish the *Revolutions*. " His hesitation . . . has often been ascribed to fear of ecclesiastical opposition." To this correct remark, Professor Africa added: " but the theme of sacerdotal oppression is difficult to maintain " (p. 403). Is the " fear of ecclesiastical opposition " identical with " sacerdotal oppression "? Copernicus felt an acute " fear of ecclesiastical opposition." But, having delayed the publication of the *Revolutions* thirty-six years, he escaped feeling " sacerdotal oppression." He died a natural death, before he could be burned at the stake, like Giordano Bruno, or sentenced to life imprisonment, like Galileo. Professor Africa introduced a further confusion in his discussion of this highly important matter by talking about Copernicus' " timidity at the professional opposition of rival astronomers " (p. 403).[13]

---

[10] *Tychonis Brahe dani opera omnia* (Copenhagen, 1913-1929), I, 149: 22-30.

[11] See pp. 326-327 of the article cited in n. 2, above.

[12] J. L. E. Dreyer, *Tycho Brahe* (Edinburgh, 1890), p. 125.

[13] If " Copernicus' attitude toward other scientists was generally . . . contempt," and if he wrote an " imperious *Letter against Wer-*

But it was not professional astronomers who burned Bruno at the stake or condemned Galileo to life imprisonment.

Professor Africa was confident that Copernicus delayed the publication of the *Revolutions* "not through fear of rack" (p. 409). But the rack was one of the most persuasive arguments in the armamentarium of the theologians. By showing Galileo the rack and other instruments of torture, the theologians persuaded him to abjure Copernicanism. And Kepler has preserved that portion of a letter to Copernicus in which his correspondent advised him to mollify the " theologians, whose opposition you fear." [14]

Copernicus began his dedication to the pope by expressing his certainty that " certain persons," as soon as they heard what his book was all about, would immediately shout that he must be hissed off the stage. Toward the close of the dedication he anticipated attacks on his book by people who knew nothing about mathematics but presumed to meddle in it " on account of some passage in Scripture." That is why Copernicus coined his single most renowned utterance: " Mathematics is written for mathematicians." He implied that it was not written for theologians who had no competence in the mathematical sciences.

Now we are in a position to understand Copernicus' conduct with regard to Aristarchus. He retained his references to Aristarchus' determinations of the obliquity of the ecliptic and the length of the year. But it was not for these strictly mathematical results that it was " thought that the Greeks ought to bring charges of impiety against Aristarchus." [15] Those charges would have been based on Aristarchus' conception of a moving earth. That was precisely what Copernicus suppressed in the deleted passage. According to

Professor Africa (p. 406), his motive in so doing was " either vanity, ' Pythagorean' scruples, or both." Actually Copernicus had no wish to inform or remind anybody that a great astronomer could have been prosecuted for impiety. If Aristarchus was indeed " the ancient Copernicus," Copernicus had no desire to become the modern Aristarchus.

It is time to delimit the sense in which Copernicus may be said to have been a Pythagorean. He was indebted to the Pythagoreans for his basic conception, the moving earth. Little else of their astronomy was known to him. What he prized was their prudence in intellectual matters.

In the letter from which Professor Africa quoted Kepler's jest about " Pythagoras, grandfather of all Copernicans," the great reformer of Copernicanism went on to advise the teacher who had introduced him to the Copernican astronomy:

Let us imitate the Pythagoreans even in their customs. If anybody approaches us privately, let us tell him our opinion frankly. Publicly let us be silent. Why go and ruin astronomy by means of astronomy? The whole world is full of men of the type who are ready to throw all of astronomy, if it sides completely with Copernicus, off the earth, and to forbid the specialists an income. But specialists cannot live off themselves or on air. Therefore let us act in astronomical affairs in such a way that we hold on to the supporters of astronomy and do not starve.[16]

The Pythagorean prudence recommended by the extrovert Kepler in a private letter exhibited more explicitly the intense fear justifiably felt by Copernicus the introvert.

What is commonly called the Copernican Revolution was not completed by Copernicus, but commenced by him. To speak, as does Professor Africa (p. 409), of " the Copernican Revolution, to which Copernicus was not party" is to perform *Hamlet* without the prince of Denmark.

---

ner," a well-known mathematician and astronomer, when did he undergo the personality change that made him feel " timidity at the professional opposition of rival astronomers "?

[14] *Three Cop. Tr.*, p. 23.
[15] Plutarch, *The Face in the Moon*, 923.

[16] *Johannes Kepler Gesammelte Werke*, XIII (Munich, 1945), 231: 505-512.

3

Copernicus' Quotation from Sophocles

Nicholas Copernicus (1473-1543) is rightfully regarded as the founder of modern astronomy because he revived and strengthened the heliocentric conception of the universe that had been propounded in classical antiquity. In his epoch-making treatise *De revolutionibus orbium coelestium* (Nuremberg 1543) he correctly contended that the earth is one of the planetary satellites of the sun, and consequently is not located in the middle of everything. In this central location, to which the earth had been erroneously assigned so long, he put the sun :

> At the midpoint of all things lies the sun. As the location of this luminary in the cosmos, that most beautiful temple, would there be any other place or any better place than the center, from which it can light up everything at the same time ? Hence the sun is not inappropriately called by some the lamp of the universe, by others its mind, and by others its ruler. Trismegistus terms it the visible god, and Sophocles' *Electra* the all-seeing (*Revolutions*, I, 10) (¹).

The last four words of Copernicus' salute to the sun, *Sophoclis Electra intuentem omnia* (*solem vocat*), present a problem to which attention was directed for the first time by Alexander von Humboldt (1769-1859). This extraordinary man was largely responsible for the founding of Berlin University, which in our day has been renamed in his honor Humboldt University by the (East) German Democratic Republic. When Humboldt was in his middle sixties, a scientist of world renown and privy councillor to the king of Prussia, he regu-

---

(¹) The holograph manuscript of Copernicus' *Revolutions* has been preserved, and a photocopy of it was published in *Nikolaus Kopernikus Gesamtausgabe* (Munich and Berlin 1944-1949) I, with our passage near the top of fol. 10r.

larly attended the lectures on Greek antiquities and literature which were given from 1833 to 1835 by the eminent philologist August Böckh (1785-1867) (²). While Humboldt was engaged in writing *Kosmos*, his most famous work, it was his practice to consult Böckh about questions pertaining to ancient authors. Thus Humboldt said that Copernicus'

> reference to Sophocles' *Electra* is confused, since in that play the sun is never explicitly termed 'all-seeing', as it is in the *Iliad* and *Odyssey* as well as in Aeschylus' *Libation-bearers* (line 980), which Copernicus would presumably not have called the *Electra*. According to Böckh's surmise, the reference is in all likelihood ascribable to a lapse of memory, and is the result of an indistinct recollection of line 869 in Sophocles' *Oedipus at Colonus* (³).

The theory suggested by Böckh (⁴) and published by Humboldt that Copernicus, trusting to his unreliable memory, unwittingly transferred Sophocles' description of the sun as all-seeing from the *Oedipus at Colonus* to the *Electra* did not long remain unchallenged. Carl Ludolf Menzzer, who translated the *Revolutions* into German, sought to defend Copernicus' quotation as it stands by calling attention to *Electra*, 823-826 :

> Where are the thunderbolts of Zeus, or
>   Where is the shining
> Sun, if, while beholding these things,
>   They hide, unmoved ?

But here the chorus chants that the sun beholds these things, these particular crimes, not all things, as Menzzer undoubtedly realized.

(²) Humboldt, quoted by FERDINAND ASCHERSON, ' August Boeckhs fünfzigjähriges Doctor-jubilaeum ', *Jahrbücher für classische Philologie* 3 *(= Neue Jahrbücher für Philologie und Paedagogik* 75 [1857] 261).

(³) HUMBOLDT, *Kosmos* (Stuttgart and Tübingen 1845-1862) 2, 500.

(⁴) Dr. Heinrich Kramm of the manuscript division of the Westdeutsche Bibliothek in Marburg informed me in a letter of July 5, 1960 that the Böckh papers in his keeping do not deal with this subject. Did Böckh make his suggestion in a conversation with Humboldt ?

Hence, aware that he could not validate Copernicus' quotation by citing only this passage, Menzzer added :

> ... especially if we connect with lines 823-826 what the chorus says to Electra in lines 174-175 :
>
> In heaven still abides mighty
> Zeus, who sees all things and governs them (⁵).

Again Menzzer's argument is faulty : the ability to see all things is here attributed to Zeus, and not to the sun. Must not our verdict consequently be that Menzzer could not find the sun described as all-seeing in the *Electra*, and that therefore his attempt to uphold Copernicus' quotation is to be judged a failure ?

Despite the decisive defect in Menzzer's position, he exerted a powerful influence. Thus Alexandre Koyré (1892-), of the Ecole Pratique des Hautes Etudes and the Institute for Advanced Study, in his French translation of the *Revolutions*, Book I, noted that 'According to C. L. Menzzer, Copernicus refers to lines 823-826 of the tragedy' (⁶). No such appeal to the authority of Menzzer was deemed necessary by John Frederic Dobson (1875-1947), professor of Greek at the University of Bristol, who declared that '*Electra*, 823-826, is the nearest approach to the reference of Copernicus' (⁷). Although lines 823-826 are the *Electra's* nearest approach to the reference of Copernicus, they cannot be the lines intended by his reference, since they do not describe the sun as all-seeing. Nevertheless Franz Zeller and Karl Zeller, the joint editors of Volume II of the sumptuous *Nikolaus Kopernikus Gesamtausgabe*, persisted in propagating Menzzer's mistake by citing *Electra* 174-175 and 823-826 without comment (⁸), as though these passages, either separately or in combination, described the sun as all-seeing, and even though the Zellers themselves admitted that Copernicus 'does not always cite previous authors correctly' (⁹).

(⁵) C. L. MENZZER, *Ueber die Kreisbewegungen der Weltkörper* (Leipzig 1879 [reprinted 1939]), p. 9, n. 33.

(⁶) A. KOYRÉ, *Nicolas Copernic, Des révolutions des orbes célestes* (Paris 1934), p. 147, n. 23.

(⁷) *De revolutionibus*, Preface and Book I, tr. J. F. DOBSON, *Royal Astronomical Society, Occasional Notes* 2 (1947), no. 10 (reprinted 1955), p. 31, n. 53.

(⁸) *Gesamtausgabe* 2, 443.

(⁹) *Op. cit.* 2, 429, lines 24-25.

The wrong turn taken by Menzzer and followed by Koyré, Dobson, and the Zellers, fortunately was reversed by Aleksander Birkenmajer (1890-   ), professor at the University of Warsaw, who quite sensibly went back to Böckh's view that Copernicus meant to cite the *Oedipus at Colonus*, 869 ([10]).

Copernicus wrote *intuentem omnia*, however, instead of quoting Sophocles' words ὁ πάντα λεύσσων "Ἥλιος, although elsewhere Copernicus evidently felt no impropriety in introducing Greek expressions into the Latin text of the *Revolutions*. In fact, in the Dedication he quoted an extended passage from *The Opinions of the Philosophers*, a Greek doxographical work then attributed to Plutarch ([1]).

When and where did Copernicus learn Greek ? At Kraków, the only university he attended in his native country, the language of ancient Hellas was not taught while he was a student there. But in Italy, the land of humanism, where he went to complete his education, the situation was quite different.

Bologna was the first Italian university to which he directed his steps. When he matriculated there in 1496, the professor of Greek, Antonio Urceo, surnamed 'Codro' (the pauper), was highly popular. To Codro, on April 17, 1499, the famous publisher Aldo Manuzio of Venice dedicated his edition of the Greek *Epistles of Various Philosophers, Orators, and Sophists* 'in order that you may show them to your pupils, who may thereby be inspired with a greater zeal to master more elegant literature'. This Aldine collection of epistolographers included ([12]) the 'Moral, Rustic, and Amatory Epistles' of Theophylactus Simocatta, a seventh-century Byzantine writer. Theophylactus' letters were translated from Greek into Latin by Copernicus, who published his version in 1509 ([13]). His close study of a part of the epistolographical collection dedicated to Codro pro-

---

([10]) Mikołaj Kopernik, *O obrotach sfer niebieskich, księga pierwsza* (ed. A. Birkenmajer [Warsaw 1953]) 114.

([11]) Available to Copernicus in the Aldine edition of *Plutarchi opuscula LXXXXII* (Venice 1509), with the quoted passage (III, 13) at p. 328 ; a convenient modern edition is Hermann Diels, *Doxographi graeci* (Berlin 1879 [re-issued 1958]) 378.

([12]) At fol. Φ 2r- Ψ 5v.

([13]) A photocopy of the original edition (Kraków 1509) was published under the title *Teofilakt Symokatta Listy tłumaczył z języka greckiego na łaciński Mikołaj Kopernik* (Warsaw 1953) ; in addition to a more legible printing of Copernicus' translation, this attractive book offered the Greek text of Theophylactus' letters as well as a Polish version of them by Jan Parandowski, who had previously published Polish renderings of Homer's *Odyssey* as well as of Longus' *Daphnis and Chloe*.

vided the basis for the hypothesis that he studied Greek at Bologna, and there learned about Theophylactus' letters ([14]).

Although there is no documentary proof that Copernicus studied Greek with Codro ([15]), a supplementary (or alternative) theory recently lost nearly all of its plausibility. More than three-quarters of a century ago Leopold Prowe (1821-1887), the author of what is still our most substantial biography of Copernicus, wrote : 'it is not improbable that Copernicus continued under Musuro the study of Greek which he had begun at Bologna' ([16]). Marco Musuro, editor of the Aldine collection of epistolographers, was appointed on July 27, 1503, as a substitute for the regular professor of Greek at the University of Padua, who had been absent for some time on a diplomatic mission for the Venetian Republic ([17]). An archival document discovered decades after Prowe's death demonstrated that Copernicus must have left Italy in that same year 1503, presumably not long after he received his doctoral law degree from the University of Ferrara on May 31 ([18]). No evidence indicates that he returned to Padua from Ferrara, or that he ever attended any classes as a post-doctoral student, so to say. In the unlikely event that he benefited in person from Musuro's instruction, the opportunity could have lasted only a few months at the very most ([19]).

While Copernicus was in Bologna, in nearby Modena the printing of an edition of Giovanni Crestone's Greek-Latin dictionary was finished on October 20, 1499 ([20]). Copernicus' well-worn copy of

---

([14]) Domenico Berti, *Copernico e le vicende del sistema copernicano in Italia* (Rome 1876) 51. Berti's hypothesis was further propagated by Carlo Malagola, *Della vita e delle opere di Antonio Urceo detto Codro* (Bologna 1878) 334-339, and by L. Prowe, *Nicolaus Coppernicus* (Berlin 1883-1884) I, part 1, 253-260.

([15]) Antonio Favaro, *Bullettino di bibliografia e di storia delle scienze matematiche e fisiche* 11 (1878) 322 ; Ezio Raimondi, *Codro e l'umanesimo a Bologna* (Bologna 1950), p. 205.

([16]) Prowe, *op. cit.*, I, pt. 1, p. 260.

([17]) Francesco Foffano, ' Marco Musuro, professore di greco a Padova ed a Venezia ', *Nuovo Archivio Veneto* 3 (1892), pt. 2, pp. 461-462.

([18]) Hans Schmauch, ' Die Rückkehr des Koppernikus aus Italien im Jahre 1503 ', *Zeitschrift für die Geschichte und Altertumskunde Ermlands* 25 (1933-1935) 229.

([19]) Prowe (*op. cit.*, I, pt. 1, pp. 319, 328-329) mistakenly believed that Copernicus remained at Padua and in Italy until 1506. Had Prowe been right, Copernicus could have studied Greek with Musuro for three years.

([20]) Despite this precise date in the colophon, L. A. Birkenmajer, *Mikołaj Kopernik* (Kraków 1900) 103, maintained that the book was not actually issued un-

Crestone still survives, its margins filled with his notes (²¹). The elementary character of these notes was emphasized by Franz Hipler (1836-1898) as an argument against the Berti-Malagola-Prowe thesis that Copernicus studied Greek under Codro at Bologna (²²). An additional argument was adduced by Ludwik Antoni Birkenmajer (1855-1929), father of Aleksander Birkenmajer (who was cited at note 10, above). Pointing out that Codro died on February 11, 1500, and insisting that Crestone was not available until after Codro's death, L. A. Birkenmajer concluded that Copernicus could not have studied Greek with Codro, and condemned the Berti-Malagola-Prowe thesis as a myth (²³).

But Copernicus was a student and teaching assistant (²⁴) at the University of Bologna throughout the last three years of Codro's professorship. Hipler's statement (²⁵) that during those three years Copernicus devoted himself exclusively to law and astronomy, with no time remaining for the study of Greek, was merely an unfounded conjecture. If Copernicus' notations in his copy of Crestone are somewhat elementary, they nevertheless show that he had already attained a level far above that of the beginner. We should be on guard against the tacit assumption that Copernicus began to study Greek only after he had acquired his copy of Crestone. Steadily improving his knowledge of Greek throughout the subsequent decade, he published his Latin translation of a part of a book which had been dedicated to Codro as reading matter suitable for the students at Bologna. This whole set of circumstances creates a strong presump-

---

til July, 1500. In his opinion, this difference of nine months was a matter of great importance, as we shall soon see.

(²¹) These notes were published most completely by L. A. BIRKENMAJER, *op. cit.* 112-115.

(²²) HIPLER, reviewing Prowe, in *Literarische Rundschau für das katholische Deutschland* 10 (1884) 177.

(²³) BIRKENMAJER, *op. cit.* 103. Birkenmajer's conclusion was recently repeated by RYSZARD GANSINIEC in the Preface to *Teofilakt Symokatta Listy*, p. XVII. I wish to express my sincere gratitude to my colleague, Dr. Alexander Groth, and to Dr. Ezri Atzmon of Hunter College, both of whom were good enough to translate excerpts from these Polish authors for me.

(²⁴) GIULIO RIGHINI, 'Copernico " Doctor Ferrariensis " e " Magister " a Bologna ', *R. Deputazione di Storia patria per l'Emilia e Romagna, sezione di Ferrara, Atti e Memorie* 1 (1942) 153.

(²⁵) *Loc. cit.*

tion, as Antonio Favaro (1847-1922) put it ([26]), that Copernicus may well have begun to learn Greek at Bologna (before the publication of Crestone), and may well have heard about Theophylactus' letters through contact (formal or informal) with Codro. We can unhesitatingly accept L. A. Birkenmajer's assertion ([27]) that Copernicus learned only a little Greek while he was at Bologna. We can with equal firmness reject the overstatement of the case by Maximilian Curtze (1837-1903), who went too far when he said that Copernicus read and mastered Theophylactus under Codro's direction ([28]). This extreme position provoked its dialectical opposite, which denies the possibility of any association between Copernicus and Codro. Somewhere between these two extremes lies the truth, patiently awaiting documentary confirmation.

Copernicus' Latin translation of Theophylactus' letters deviates considerably from the Greek text as printed for the first time in the Aldine collection of epistolographers. In order to account for these deviations, Hipler suggested that Copernicus' translation was apparently based in part on a manuscript whose readings differed from the Aldine ([29]). Hipler's suggestion was uncritically transformed into a 'fact' by Prowe ([30]), and was at first accepted even by Theodor Nissen, author of our most painstaking philological analysis of Copernicus' translation of Theophylactus ([31]). Later, when Nissen's attention was directed to L. A. Birkenmajer's remark ([32]) that it is unnecessary to posit the existence of a Greek manuscript of Theophylactus' letters in Copernicus' hands, Nissen explicitly abandoned the Hipler-Prowe thesis ([33]). But Nissen clutched at the possibility of a printed Theophylactus edition separate from the Aldine and dif-

---

([26]) *Loc. cit.*

([27]) *Op. cit.*, 101.

([28]) Curtze, ' Nicolaus Coppernicus, ' *Himmel und Erde* 11 (1899) 264. Prowe (*op. cit.* I, pt. 1, pp. 399-400) had previously pointed to Copernicus' translation as proof that Codro expounded Theophylactus to his pupils.

([29]) Hipler, *Spicilegium copernicanum* (Braunsberg 1873) 73.

([30]) *Op. cit.*, 2, 45.

([31]) Nissen, ' Die Briefe des Theophylaktos Simokattes und ihre Lateinische Uebersetzung durch Nikolaus Coppernicus ', *Byzantinisch-neugriechische Jahrbücher* 13(1936-1937) 41.

([32]) *Op. cit.* 123.

([33]) Nissen, ' Zur Theophylakt-Uebersetzung des Coppernicus ', *Byz.-neugr. Jahrbücher* 14 (1937-1938) 41.

fering from it, as the explanation of Copernicus' deviations from the Aldine text.

No such separate and different printed edition of Theophylactus has been found. Nor is any Greek manuscript of Theophylactus' letters known to have been accessible to Copernicus ; none of the extant manuscripts agrees with his deviations from the Aldine text. He omits material present in the Aldine. He never supplies material missing in the Aldine ; in fact, he fails to recognize its absence. In short, Copernicus based his Latin translation on the Aldine edition, which offered the first printed Greek text of Theophylactus' letters. Nothing indicates that any other printed text, or any manuscript, was available to him.

How, then, are his deviations from the Aldine text to be explained ? Some of its readings make no sense [34] ; later translators, wishing to make sense, had to adopt the same expedients as Copernicus, namely, deviations or omissions from the Aldine text. Where it does make sense, Copernicus did not always apprehend the sense. Many of the words used by Theophylactus in his straining for an elegant style were not to be found in Crestone's pitifully meager dictionary, the only one available to Copernicus. He was not a Greek scholar of the highest rank [35], like his contemporary, Desiderius Erasmus. Although he was indisputably a humanist, his greatest contribution to the enlightenment of mankind lay outside the domain of classical philology. His deviations from the Aldine edition are ascribable, then, in no small measure to its defects as a Greek text, and for the rest to his shortcomings as a Greek scholar [36].

Copernicus' career as a Greek scholar may be tentatively summarized as follows : he began to study the language at Bologna ; there, probably through Codro, he heard about Theophylactus [37] ;

---

[34] E.g., NISSEN, *op. cit.* 13(1936-1937) pp. 42, 43, 48.

[35] His deficiencies as a student of Greek were emphasized by JERZY KOWALSKI, ' Kopernik jako filolog i pisarz łacinski ', in *Mikołaj Kopernik* (Lvov and Warsaw 1924) 132-141 ; Kowalski imagined that Copernicus worked from an inferior manuscript.

[36] Few, if any, of Copernicus' deviations from the Aldine edition evince the kind of erudition required to emend a classical text. Yet CURTZE (*op. cit.* 203) "explained" Copernicus' deviations as variant readings proposed by Codro, a veteran professor of Greek.

[37] IVAN IVANOVICH TOLSTOI, 'Copernicus and his Latin Translation of Theophylactus Simocatta's Letters', in *Nikolai Kopernik* (ed. Aleksandr A. Mikhailov

he increased his knowledge with the help of Crestone ; by self-im-
provement, after his return to his native land, he reached the point
where he could publish his Latin version. This slender volume was
the first independent translation of a Greek author to be printed in
Poland. It constituted Copernicus' modest contribution to the spread
of the humanist movement in his native region.

Besides Theophylactus' eighty-five brief letters, all fictitious,
and either exchanged by imaginary persons or playfully attributed
with nonchalant disregard of chronology to famous historical figures,
Copernicus translated into Latin another sham epistle in the Aldine
collection (³⁸). This supposed communication from Lysis the Pythagorean to Hipparchus was originally inserted in the *Revolutions* by
Copernicus, but on later reflection he deleted it (³⁹).

What we have seen of Copernicus' communings with Greek authors should suffice to convince us that his decision to write *intuentem
omnia* (⁴⁰) instead of quoting Sophocles directly was not due to lack
of familiarity with the dramatist's language. Let us now look at
another of Copernicus' references to an ancient Greek author. In the
introduction to Book V of his *Revolutions*, which is devoted to the
five planets, Copernicus said :

> In Plato's *Timaeus* these five planets are each named in
> accordance with its appearance : Saturn, 'Phainon',
> as though you should say 'conspicuous' or 'visible',
> since it is hidden less than the others, and emerges more
> quickly after being obscured by the sun ; Jupiter, 'Phae-
> thon', from its brilliance ; Mars, 'Pyroeis', from its
> fiery splendor ; Venus, sometimes 'Phosphorus', some-
> times 'Hesperus', that is, Morning Star and Evening

[Moscow and Leningrad 1947]) 76, stressed the high cost of printed classical texts (I
wish to thank Edythe Lutzker, M.A., for her kindness in translating Tolstoi's article
from Russian for me). In 1499, the year in which the Aldine collection of epistolographers was published, Copernicus happened to be in financial straits and had to borrow money (PROWE, I, pt. I, pp. 265-267). Did he perhaps, being unable to buy the
Aldine printed text, copy a part of it by hand for his own use ? Is this the reason
why no hint has ever been found that he owned a copy of the Aldine printed edition ?

(³⁸) Fol. Γ 6v-7v.

(³⁹) Copernicus' translation is printed in *Gesamtausgabe* 2, 30-31.

(⁴⁰) Copernicus was echoing Pliny, *Natural History* 2, 6, 13 : sol omnia intuens.
For Copernicus' use of Pliny, see EDWARD ROSEN, *Three Copernican Treatises*, 2d
ed. (New York, Dover Publications 1959) 78.

Star, depending on whether it shines at dawn or at twilight ; finally, Mercury, 'Stilbon', from its vibrant and glittering light.

But in the *Timaeus* Plato did not call the five planets by these names, which were introduced long after his death ([41]). In the *Timaeus* (38D) he referred to Venus as Heosphorus (Dawn-bringer, the traditional term from Homer on), and to Mercury as 'the body called sacred to Hermes'. With these two inner planets he grouped the three outer planets simply as 'five other heavenly bodies, having the name "planets" ' (38C).

At note 11, above, we saw that Copernicus quoted a passage from (pseudo-) Plutarch's *Opinions of the Philosophers*. That same work (II, 15: Concerning the Arrangement of the Heavenly Bodies) declared :

After giving the location of the fixed stars, Plato put first Phainon, called the planet of Saturn ; second, Phaethon, the planet of Jupiter ; third, Pyroeis, the planet of Mars ; fourth, Heosphorus, the planet of Venus ; fifth, Stilbon, the planet of Mercury ([42]).

From the occurrence of Phainon, Phaethon, Pyroeis, and Stilbon in pseudo-Plutarch's statement about Plato's arrangement of the heavenly bodies, did Copernicus incorrectly infer that these planetary designations were used by Plato himself ([43]) ? And did Copernicus then gratuitously assign these designations to Plato's cosmological dialogue, the *Timaeus* ?

If we have correctly reconstructed the two steps which misled Copernicus into attributing these post-Platonic planetary designations to the *Timaeus*, it is clear that he never took the trouble to re-read that dialogue in order to test his double assumption. By the same token he never re-read Sophocles for the purpose of verifying in what play the tragedian described the sun as all-seeing. Even

---

([41]) FRANZ CUMONT, ' Les noms des planètes et l'astrolâtrie chez les Grecs ', *L'antiquité classique* 4 (1935) 19-24.

([42]) *Ed.* 1509, p. 321 ; DIELS, *Dox. gr.*, pp. 344-345.

([43]) As CUMONT pointed out (*Op. cit.*, pp. 6, 12, 30), ps.-Plutarch used the planetary nomenclature current in his own time (whenever that was), and he did not use Plato's nomenclature in his summary of Plato's arrangement of the planets.

today, when the process of verification has been facilitated by the publication of aids which did not exist in Copernicus' time, like the Ellendt-Genthe *Lexicon Sophocleum* (re-issued, Hildesheim 1958), writers do not always bother to check their references. This neglect of a modern scholar's duty may produce a serious distortion of history [44]. But no such duty was acknowledged four hundred years ago.

Anybody who suffers a lapse of memory will be unaware of that fact if he feels certain that his recollection is sound. In that state of mind he will be disinclined to undertake a lot of tedious checking. Those who really understand Copernicus' personality as a scientist (such understanding is utterly lacking in the most recent writer on that subject) [45] do not regard him as intellectually arrogant or indolent or likely to make a deliberate misstatement. His misquotation of Sophocles' *Electra*, like his blunder concerning Plato's *Timaeus*, should be ascribed to faulty memory or unquestioning acceptance of supposedly reliable 'authorities' in an age when thorough checking of references had not yet been established as a moral obligation of a conscientious scholar.

---

[44] One such distortion is rectified by EDWARD ROSEN, 'Calvin's Attitude toward Copernicus', below, 161–72.

[45] ARTHUR KOESTLER, *The Sleepwalkers* (London and New York 1959).

4

The Alfonsine Tables and Copernicus
(with the Assistance of Erna Hilfstein)

I THE ORIGIN AND SPREAD OF THE *ALFONSINE TABLES*

Alfonso X, king of Castile and León from 1252 to 1284, is known as "the Learned" (el Sabio). His numerous intellectual achievements include the *Alfonsine Tables,* which depended on an earlier treatise in Arabic. This was translated, not into Latin, but into Castilian Spanish "in order that the people should understand it better."[1] This policy was virtually a new departure, since a scientific treatise in Castilian was quite a rarity before Alfonso X. Although the Arabic prototype served as a model, its defects were recognized and rectified. This purpose was emphasized in the Preface preceding the Introduction to the *Alfonsine Tables:*

> Judah, son of Moses, son of Mosca,[2] and Rabbi Isaac ibn Sid say: The science of astronomy is a subject that can be investigated only through observations. The observations made by the experts devoted to this subject cannot be completed by a single man, because they cannot be finished in the lifetime of one man. On the contrary, when a result is attained, it is

---

[1] *Lapidario del rey D. Alfonso X,* ed. José Fernández Montaña (Madrid, 1881), p. 1, a succinct expression of El Sabio's linguistic program.
[2] This form of double patronymic ("Yhuda. fi de Mose. fi de Mosca.") was sometimes used by medieval Jews. The paternal grandfather of this Judah ben Moses was named (or perhaps nicknamed) Mosca ("Fly"). The grandson is called in the *Lapidario* (p. 1) "Yhuda Mosca el menor," presumably in order to distinguish him as "Mosca junior" from the grandfather "Mosca senior." Judah was ordered to translate the *Lapidario* from Arabic into Castilian by Don Alfonso, while the latter was the heir apparent, and "the translation was finished in the second year after his father, the noble king Don Fernando, conquered the city of Seville" in 1248. According to the *Lapidario,* Judah "was well informed in the art of astronomy." For that reason he was later chosen by King Alfonso X to be a co-author of the *Alfonsine Tables.* In two different Latin translations of the *Liber de iudiciis astrologiae* he is called "Judah filius muce" and "iuda filius mosse," as quoted by Evelyn S. Procter, "The Scientific Works of the Court of Alfonso X of Castile: the King and his Collaborators," *Modern Language Review,* 40 (1945), 20. "Iuda filius Mosse Alchoen" (Judah ben Moses Cohen) is said to have finished translating al-Zarqali's *Saphea* into Latin in 1231 after six years of work, according to MS 10,053, National Library, Madrid, as quoted by José María Millás Vallicrosa, *Estudios sobre Azarquiel* (Madrid/Granada, 1943-1950), p. 452. Because MS 10,053 is a thirteenth-century document, the form in which it gives Judah's name is valuable, despite the implausibility of its attribution of the *Saphea* translation to him.

attained through the work of many men, laboring one after another for a long time. The reason for this is that in the movements of the heavens there are some motions that are so slow that they complete a circuit only after thousands of years. It is therefore necessary to continue the observations because, as they are continued, phenomena will be visible at one time which were not visible at another time.

We are now in our time in the first decade of the fourth century of the second millennium of the era of Caesar.[3] Since al-Zarqali's observations two hundred years have passed.[4] In some of the positions which he adopted,[5] there have appeared changes which are obvious and manifest to the senses so that no excuse can be offered for [retaining] them. At this time there appeared the happy reign, favored by God, and the rule of the very exalted and most noble lord Don Alonso, may God preserve him! Because he loved learned men and appreciated them, he had the instruments made which are mentioned by Ptolemy in the *Almagest,* such as the armillary sphere and other devices. And he bade us observe in the city of Toledo, which is one of the principal cities of Spain, may God preserve it! In it the observations of al-Zarqali were made.[6]

Don Alonso ordered the improvement and correction of the divergences and disagreements which appeared in some places of some of the planets and in other movements. We obeyed his order, as we had to. We reconstructed the instruments as well as can be done. We worked at our observations a certain length of time and proceeded to observe the sun throughout a complete year. Before that time and thereafter we observed it whenever it entered the equinoxes and the solstices and the other quadrants of the heaven, which are the middle of the Bull, Scorpion, Lion, and Water Bearer. We also observed some conjunctions of the planets when they meet one another,

---

[3] In Spain, the era of Caesar began on 1 January 38 B.C. Hence, by this cumbersome dating of the composition of their Preface, the co-authors of the *Alfonsine Tables* placed their work within the ten years 1301-1310 of the era of Caesar, or 1263-1272 of the Christian era.

[4] Al-Zarqali performed his last observations in 1088, and died on 15 October 1100, according to the collection of biographies of Spanish Muslim scholars compiled by Ibn al-Abbar (1199-1260), as cited by Millás, *Azarquiel,* p. 10.

[5] In saying "en algunas de las posturas que él pusso" the co-authors of this Preface to the *Alfonsine Tables* would seem to be referring to the *Toledan Tables,* and to be assuming that these were compiled by al-Zarqali. The Arabic text of the *Toledan Tables* has not been recovered, but the canons to these *Tables* were ascribed, in the Latin translation, to al-Zarqali and so, by extension, the *Toledan Tables* too were attributed to him; see Millás, *Azarquiel,* pp. 18, 36, 58-59. The relation of the *Toledan Tables* to earlier tables was examined by G. J. Toomer, "A Survey of the Toledan Tables," *Osiris,* 15, (1968), 5-174.

[6] The name, or rather nickname, of this Spanish Muslim astronomer is now derived from the Arabic definite article al; zarqa, the Arabic adjective meaning "blue," used in the feminine form, because the Arabic word for "eye" is feminine; and, finally, some remnant of the Spanish diminutive suffix. Hence, this hybrid nickname would designate "the man with the little blue" eyes (according to Juan Vernet Ginés' forthcoming article on al-Zarqali in the *Dictionary of Scientific Biography,* cited hereafter as DSB).

and their conjunctions when they met some of the fixed stars. We observed many eclipses of the sun and moon, and we made other observations in which we were uncertain and we repeated them many times to resolve the doubt. We did not fail to look for and investigate anything until we saw the correction of that which had to be corrected.

Having made a complete examination, we have accepted as verified what is certain or nearly so. We made these tables on radices which are taken from those observations. We have added to them the chapters which seemed to us to be needed in this work and have called this book, the book of the *Alfonsine Tables,* because it was made and compiled at his behest. We have divided it into 54 Chapters, which follow.

These 54 Chapters, together with the preceding Preface, were printed for the first time, almost exactly six centuries after they were written, from a unique manuscript preserved in the National Library of Madrid.[7] But the Tables themselves, to which these 54 Chapters served as an Introduction, have never been printed. Indeed, no manuscript of these Tables has ever been found.

After the *Alfonsine Tables* were released in the course of the decade 1263-1272 in Toledo they drop out of sight until they reappear in Paris and then in Oxford. When did they reach the French capital? There is no space here to examine that disputed question, but the earliest date supported by satisfactory evidence is 1320. "At what date the *Alfonsine Tables* first became known at Oxford is uncertain; but it must have been after 1320, since that year is the first one occurring in any Oxford codex."[8] If the *Alfonsine Tables* reached Paris in 1320 and Oxford thereafter, a veil of obscurity still hides the half-century between their promulgation in Toledo and their first appearance in Paris and Oxford. This half-century may be called the lost years of the *Alfonsine Tables.*

In Paris the *Alfonsine Tables* underwent a radical transformation. In the first place, in their original Castilian form the (lost) *Alfonsine Tables* presumably used Roman numerals. This presumption may reasonably be based on the employment of Roman numerals throughout the other translations and tables prepared under the auspices of Alfonso X.[9] By contrast, the Parisian revision of the

---

[7] Manuel Rico y Sinobas, ed., *Libros del saber de astronomía* (Madrid, 1863-1867), IV, 111-183.

[8] J.L.E. Dreyer, "On the Original Form of the Alfonsine Tables," *Monthly Notices of the Royal Astronomical Society,* London, 80 (1920), 243-262; 252: Bodleian Laud Misc. 594.

[9] The apparent exceptions (Rico, III, 53, 114: tables of longitudes, latitudes, and longest and shortest days of the year for various cities) were inserted by Rico to fill areas left blank by Alfonso X's astronomers.

Alfonsine Tables shifted to Arabic numerals.[10] In Volume IV of the *Libros del Saber de Astronomía del Rey D. Alfonso X de Castilla* the editor introduced material which he labeled "Numerical Fragments of the Alfonsine Tables." He took these fragments from two manuscripts, one written in 1309 and the other in 1396. Apart from their late date, these fragments would not have belonged to the original *Alfonsine Tables* of 1263-1272, since those used Roman numerals whereas the fragments employ Arabic numerals.

A second striking difference between the original form of the *Alfonsine Tables* and the Parisian revision of them concerns the definition of an astronomical sign. According to Chapter 15 of the Introduction to the Spanish *Tables,* a sign consists of XXX degrees.[11] But this conventional concept of a sign (the common sign, as it came to be called) was replaced in the Parisian version by a "physical sign" of 60°.[12] This division of a circle into 6 signs of 60° each, instead of 12 signs of 30° each, exemplifies the Parisian preference for a sexagesimal system. This shift to a sweeping sexagesimal system made it much easier to use the tabular material. This motive for the Parisian transformation of the *Alfonsine Tables* operated also on the other side of the English Channel. A similar spirit moved John of Saxony to compose his *Almanac Compiled according to the* Alfonsine Tables *for the Years 1336-1380 and Reduced to the Meridian of Paris.*

It was John of Saxony's 1327 Canons for the Parisian version of the *Alfonsine Tables* that achieved the greatest popularity. His Canons are preserved in numerous manuscripts as well as in the first printed edition of the *Alfonsine Tables* (Venice, 1483). By contrast, the tabular material of the original *Alfonsine Tables* in Spanish has not been recovered, and the accompanying Introduction barely survived in a single manuscript. El Sabio's choice of the vernacular as the vehicle for the popularization of knowledge

---

[10] Edward Grant, ed., *A Source Book in Medieval Science* (Harvard University Press, 1974), pp. 479-489, reproduces tabular material expressed in Arabic numerals and taken from the first printed edition of the Parisian version of the *Alfonsine Tables.*

[11] Rico, IV, 133.

[12] John of Saxony's canons, for example, contained the following explanation: "In using the *Alfonsine Tables* we proceed in a physical manner, that is, with sexagesimal numbers. Now in astronomy the degrees are called the 'units.' And when 60 degrees have been accumulated, in their place in these tables is put one sign. For, a degree is divided into 60 minutes, a minute into 60 seconds, and so on" (*Alfonsine Tables,* Venice, 1483, sig. a2r; cf. Grant, *Source Book,* pp. 466-467).

limited the usefulness of his *Alfonsine Tables* to the Iberian peninsula or a portion thereof, whereas the Paris school's preference for Latin made their version of the *Alfonsine Tables* accessible to educated circles throughout the Latin West.

II COPERNICUS' ACQUISITION OF A COPY OF THE *ALFONSINE TABLES*

When Nicholas Copernicus (1473-1543), the founder of modern astronomy, was a mature man, the Fifth Lateran Council (1512-1517) attempted to wrestle with the problem of the calendar. Copernicus played only a small part in that discussion because he felt that the heavenly motions defining the month and the year were not yet understood with sufficient precision.

The first printed edition (Venice, 1483) of the *Alfonsine Tables,* containing John of Saxony's Canons as well as the Parisian transformation of the tables themselves, had appeared when Copernicus was only ten years old. But the second edition (Venice, 1492) was issued almost exactly a year after Copernicus had entered the University of Cracow. Even in his late teens he was already interested in mathematics and astronomy, as is shown by his acquiring a copy of Euclid's *Elements* (Venice, 1482; the first edition of the Latin translation) as well as a copy of the second edition of the *Alfonsine Tables.*[13] Copernicus' copy of Euclid and his copy of the *Alfonsine Tables* are both preserved today in the library of the University of Uppsala, Sweden. Each is bound with another work, the Euclid with an astrological treatise (Venice, 1485), and the *Alfonsine Tables* with Regiomontanus' *Tables of Directions* (Augsburg, 1490). In each case the binding consists of wooden boards covered with embossed leather, front and back.[14] In their ornamentation, these two bindings resemble each other as well as the bindings of other incunabula in the Jagiellonian Library of the University of Cracow. All these bindings appear to be the handiwork of local craftsmen. It therefore seems entirely likely that Copernicus, as a young as-

---

[13] The title page of the *Alfonsine Tables* showing Copernicus' ownership signature is reproduced on a plate between pages 28 and 29 in Ludwik Antoni Birkenmajer, *Mikołaj Kopernik* (Cracow, 1900). An English version of this work is now available (Xerox University Microfilms, LD00044).

[14] Photographs of the front and rear bindings of the *Alfonsine Tables* are presented as two plates between pages 26 and 27 in Birkenmajer, *Kopernik.* In the legend below the second plate, "late XVIth-century" appears by a misprint for "late XVth-century" as the date of the binding. Similar bindings of other incunabula are listed on p. 679 of the 1900 edition, as well as in L.A. Birkenmajer, *Stromata Copernicana* (Cracow, 1924), pp. 338-342.

tronomy buff at the University of Cracow who enjoyed the moral and financial support of his uncle, the bishop of Varmia, bought a copy of the *Alfonsine Tables*, wrote his name on its title page to indicate ownership, and had it bound in Cracow.

The authentic Copernicus copy of the *Alfonsine Tables* is now in Uppsala, as we just saw. It is a printed book, an example of the second edition (Venice, 1492). By contrast, a fifteenth-century manuscript copy of the *Alfonsine Tables*, held by the Mediceo-Laurentian Library in Florence as Codex Ashburnham 1697, purports to carry Copernicus' signature as owner.[15] The fraudulent character of this signature was recently exposed.[16] The notorious forger-thief-scholar Guglielmo Libri (1803-1869) imitated a genuine Copernicus signature occurring in a letter that was published for the first time in 1843. In signing this letter, Copernicus naturally used his name in the nominative case. Hence, both his given name and his surname end with a downward sweeping stroke that unmistakably belongs to a final -s.[17] More suitable for an ownership signature, however, is the genitive case. Accordingly, excellent paleographer that he was, Libri in his imitation replaced the terminal -s of the letter with the possessive -i of the forgery.[18] The market price of the manuscript thus considerably enhanced, Libri included this copy of the *Alfonsine Tables* in the collection which he sold in 1847 to Lord Ashburnham, from whom it was bought in 1884 by the Italian government for the library in Florence. As a result, an otherwise authentic manuscript copy of the *Alfonsine Tables* in Italy bears a forged ownership signature of Copernicus, who did possess that printed copy of the *Alfonsine Tables* which is now in Sweden.

That copy of the *Alfonsine Tables* was bound, as we just saw, with Copernicus' copy of Regiomontanus' *Tables of Directions*. "These two books together provide the working library for an astronomer of Copernicus' time—the *Tabulae directionum* for spherical astronomy, the *Alfonsine Tables* for planetary theory and eclipses"; "the *Alphonsine Tables* . . . is . . . the source of almost every parameter in the *Commentariolus*," according to Professor

---

[15] Paul Oskar Kristeller, *Iter italicum* (London, 1963-1967), I, 87.

[16] Jerzy Dobrzycki, "Florencki rzekomy autograf Mikołaja Kopernika," *Kwartalnik Historii Nauki i Techniki*, 12 (1967), 291-293.

[17] Dobrzycki, Plate 3, between pages 292 and 293.

[18] Dobrzycki, Plate 2, between pages 292 and 293.

Noel M. Swerdlow, who recently edited and translated Copernicus' *Commentariolus*.[19]

In the *Commentariolus* Copernicus said that the planets' "apsides [are] immovable: Saturn's near the star described as being above the Archer's elbow" (. . . absides suas invariabiles, Saturnus quidem circa stellam quae super cubitum esse dicitur Sagittatoris). By saying that Saturn's higher apse (or apogee) is near this star in the Archer, Copernicus means that the planet's apogee has almost the same celestial longitude as this star. This star is described as "Que est super cubitum dextrum" in the constellation "Sagictarii," in Copernicus' edition of the *Alfonsine Tables*, which gives the star's celestial longitude as 4 signs, 41°58' (sig. c4r). Two kinds of signs were distinguished by the editor of the 1492 edition, Johannes Lucilius Santritter of Heilbronn, in his first Canon or Proposition:

> A degree . . . is the sixtieth part of a physical sign, six of which make a circle or revolution; or [a degree is] the thirtieth part of a common sign, twelve of which make a circle or revolution. . . . Hence, when 60° have accumulated, they are replaced in these *Tables* more often by one physical sign, although in some of the *Tables* included herein, if 30° have accumulated, they are replaced by one common sign [signum commune], as will be clear to whoever uses [these *Tables*, sig. A5r].[20]

In the case of the Archer's elbow, the physical sign is intended rather than the common sign, so that our star's celestial longitude = 4 × 60° (=240°) + 41°58' = 281°58'.

The modern designation of the Archer's elbow is "h^2 Sagittarii." Using the notation "Lambda$_A$ Saturn" for the "longitude of Saturn's apogee" Swerdlow says (pp. 479-480):

> We now compare the tropical longitudes of the apogees from the *Alphonsine Tables* for the supposed epoch of the Alphonsine star catalog, Era Alfonso = June 0, 1252, with the longitudes of these stars in the catalog.
>
> h^2 Sagittarii—Lambda$_A$ Saturn = 4,41;58°−4,10;39° = 31;19° west

Swerdlow's "4,41;58°" = 4 signs, 41°58' = 281°58' = the longitude of the Archer's elbow in the Alfonsine star catalog (1492 ed., sig. c4r, line 6). By the same token Swerdlow's "4,10;39°" = 4 signs, 10°39' = 250°39' = the Alfonsine longitude of Saturn's apogee (according to Swerdlow, who does not indicate his source

---

[19] Noel M. Swerdlow, "The Derivation and First Draft of Copernicus' Planetary Theory: A Translation of the Commentariolus with Commentary," *Proceedings of the American Philosophical Society*, 117 (1973), 423-512; pp. 427, 452.

[20] "John of Lignères used the physical signs of thirty degrees and not the natural signs of sixty degrees," according to DSB, VII, 123-124. If so, John of Lignères' physical sign =30° was transformed by his pupil John of Saxony into the physical sign =60°, the definition retained by Santritter.

for this parameter). In Copernicus' 1492 edition of the *Alfonsine Tables* (sig. b1v), the table entitled "Radices of Saturn's apogee at the eras given here, without the motion of the eighth sphere" locates Saturn's apogee for the Era of Alfonso at 4 2 35 20 41 0 = 4 signs, $2°35'20''41'''0''''$ = $242°35'$ (without the negligible smaller fractions). Hence Saturn's apogee is nearly 40° west of the Archer's elbow ($281°58'$ - $242°35'$ = $39°23'$, not "31;19°," Swerdlow's notation for $31°19'$).

Whether the longitudinal separation between the Archer's elbow and Saturn's apogee is nearly 40° (according to Copernicus' edition of the *Alfonsine Tables*) or $31°19'$ (according to Swerdlow), according to Copernicus' *Commentariolus* there is very little longitudinal separation between the Archer's elbow and Saturn's apogee. This parameter, therefore, was not obtained by Copernicus from the *Alfonsine Tables*. Instead of drawing this conclusion, Swerdlow insists that Copernicus' *Commentariolus* located Saturn's apogee near the Alfonsine Archer's elbow = $281°58'$. "Since such a longitude for Saturn's apogee is unheard of," he remarks (p. 480), "I would guess that Copernicus made an error of 1^s = 30° in noting its position." But the discrepancy is nearly 40°, not 30°, and in this context one sign (1^s) = 60°, not 30°.

Rather than guess that Copernicus made an error, let us see where he placed Saturn's apogee at two stages of his professional development. In 1527, at the peak of his career, Copernicus found Saturn's apogee at $240°21'$ (*Revolutions*, V, 6), the value he recorded in the notes he wrote in his copy of the *Alfonsine Tables* (reproduced by Swerdlow, p. 427). By 1527 Copernicus had discovered the gradual shift of the planetary apsidal lines with respect to the stars.[21] Before making that discovery, he had put Saturn's apogee at $226°30'$ (in his star catalog, *Revolutions*, II, 14).

In antiquity Ptolemy's *Syntaxis* (long miscalled the *Almagest;* XI, 8) put Saturn's apogee at $224°10'$. In the *Epitome* (XI, 14) of Ptolemy's *Syntaxis* by Peurbach and Regiomontanus (ed. Venice, 1496, sig. n1r), Saturn's apogee was moved eastward to $233°46'$, and to $240°5'$ in the Toledan Tables[22] and in Campanus of Novara.[23] The latter's *Theory of the Planets* was copied at Cracow in 1494, when Copernicus was a student there, and a Cracow

---

[21] H. Hugonnard-Roche, E. Rosen, and J.-P. Verdet, *Introductions à l'astronomie de Copernic* (Paris, 1975), pp. 35, 108-109.

[22] Toomer, *Osiris*, 15 (1968) 45.

[23] Francis S. Benjamin and G. J. Toomer, *Campanus of Novara and Medieval Planetary Theory* (Madison, 1971), p. 301.

annotation put Saturn's apogee at 252°45'25" in 1473, the year in which Copernicus was born.[24] He wrote the *Commentariolus* about 1510. By that time the steadily increasing values for the longitude of Saturn's apogee may have brought that parameter somewhere near 264°50', Ptolemy's longitude for the Archer's elbow, available to Copernicus in Giorgio Valla's *De expetendis et fugiendis rebus* (Venice, 1501, sig. ee1r). Valla's Latin translations of ancient Greek mathematical and astronomical works were enormously valuable to Copernicus, since the originals were virtually inaccessible, and Copernicus' knowledge of Greek was still rather rudimentary. But Valla's value for Saturn's apogee was 230°10' (sig. gg1v, lines 6-4 up), at a distance of 34°40' from the Archer's elbow = 264°50'. Clearly, Valla was not the sole reason why the *Commentariolus* put Saturn's apogee near the Archer's elbow. Copernicus may have taken this star's longitude from Valla, but the planet's apogee from some other (still unidentified) source. It was not Copernicus' habit to specify his sources in any systematic way.

Swerdlow's claim (p. 452) that "the *Alphonsine Tables* . . . is . . . the source of almost every parameter in the *Commentariolus*" is not supported by any documentation.[25] The *Alfonsine Tables* were taught at the University of Cracow for the last time in the summer semester of 1488.[26] In the following winter semester (1488-1489) they were replaced by the simplified tables called *Tabulae resolutae*.[27] This replacement is marked by a characteristically human error in the official archives: at first the subject was listed as the *Alfonsine Tables* (presumably by force of habit), but the word *resolutas* was promptly substituted for *Alphoncij*.[28] Then in the summer semester of 1489 the *Tabulae resolutae* were taught once

---

[24] *Ibid.*

[25] The *Commentariolus* locates Mars' ascending node at "6½° west of Vergiliae." According to Swerdlow (p. 488), "the *Vergiliae* are the Pleiades, and it is not certain which of the four stars in the group listed in the Alphonsine catalog Copernicus intends by this vague designation." In Copernicus' *Alfonsine Tables*, however, no star is called Vergiliae. But in his own star catalog Copernicus wrote "Vergiliae" in the right margin alongside Taurus 30 (Nicholas Copernicus, *Complete Works*, I: *The Manuscript of Nicholas Copernicus' On the Revolutions, Facsimile*, London, 1972, fol. 59r). For Copernicus, Vergiliae was a precise, not a vague, designation, which he did not obtain from his *Alfonsine Tables*.

[26] *Liber diligentiarum facultatis artisticae universitatis Cracoviensis*, ed. W. Wisłocki (Cracow, 1886), p. 4.

[27] Ernst Zinner, *Entstehung und Ausbreitung der coppernicanischen Lehre* (Erlangen, 1943), pp. 503-504; *Studia Copernicana*, IV (Wrocław, 1972), 518.

[28] *Liber diligentiarum*, p. 9.

more before Copernicus entered the University of Cracow in the
winter semester of 1491-1492. While he was a student at Cracow,
the *Tabulae resolutae* were taught in the winter semesters of 1492-
1493 and 1493-1494 as well as in the summer semester of 1494
(and of 1495, if he was still there.).[29] Presumably it was on one of
these three (or four) occasions that he acquired a copy of the second
edition of the *Alfonsine Tables,* completed in Venice on 31 October
1492. This is the only one of Copernicus' books in which he ordered a gathering of blank pages to be bound together with the
printed work. Like the other students of astronomy at the University of Cracow, Copernicus was aware that the *Alfonsine Tables*
had recently been superseded by the *Tabulae resolutae.* These,
however, were not yet available in printed form. Hence, the sixteen
blank pages would provide Copernicus with a convenient place to
record improvements in the *Alfonsine Tables.*

It was from the *Alfonsine Tables,* according to Swerdlow (p.
452), that the *Commentariolus* took "365 days, 5 hours, 49
minutes" as the length of the tropical year which Copernicus
attributed to "Hispalensis." But Copernicus' Hispalensis has been
shown to be Alfonso de Corduba Hispalensis, whose *Almanach
perpetuum* (sig. a1v) gave the length of the year as "365¼ days
minus 11 minutes of an hour" = $365^d6^h - 11^m = 365^d5^h49^m$.
Hispalensis' *Almanach* was published in Venice on 15 July 1502,
when Copernicus was residing in Padua, some twenty miles from
the place of publication. Copernicus was then enrolled as a student
at the University of Padua, which was under the control of Venice.
Intensely absorbed in astronomy, Copernicus naturally learned of
the nearby publication of Hispalensis' *Almanach.* From it he took
the length of the year as $365^d5^h49^m$. This length he attributed to
Hispalensis.by name in his *Commentariolus,* which he wrote a few
years later.

The foregoing simple and straightforward explanation of Copernicus' reference to Hispalensis' $365^d5^h49^m$ has commended itself to
every student of the subject in the half-century since the Hispalensis
problem was solved.[30] This correct solution of an old, nagging

---

[29] *Ibid.,* pp. 22, 25, 28, 31.

[30] L. A. Birkenmajer, *Stromata Copernicana,* p. 353. Hispalensis' *Tabule astro-
nomice Elisabeth regine* was published in Venice on 28 December 1503, and
again in 1517. A copy of either one of these two editions of Hispalensis'
Astronomical Tables was bound with a copy of the *Alfonsine Tables,* according
to the 1563 inventory of Regiomontanus' personal library, as augmented after
his death (Ernst Zinner, *Leben und Wirken des Joh. Müller von Königsberg*

puzzle was recently challenged, however, by Swerdlow (p. 452), who asked: "Could *Hispalensis* in fact be a misreading of *Hispaniensis*?" But all three extant manuscripts of the *Commentariolus* read "Hispalensis." Without specifying who could have been guilty of the supposed misreading, Swerdlow was confident that *"Hispaniensis . . .* would undoubtedly refer to good king Alfonso," since "Alfonso X . . . was, after all, a Spaniard." An ordinary subject of King Alfonso X might curtly be called *Hispaniensis.* But Copernicus' *Alfonsine Tables* mentioned their royal and imperial patron four times: once, as "king of Castile," and three times as "king of the Romans and of Castile," referring to Alfonso's partially successful candidacy in the inconclusive election of the Holy Roman Emperor in 1257 (sig. A4r, B1r, a1r, k6r). In any case, Copernicus' term for a Spaniard was *Hispanus,* not *Hispaniensis* (*Revolutions,* III, 2, 6, 16.).

Swerdlow also argued that Hispalensis' $365^d5^h49^m$ "is simply a rounding of the tropical year in the *Alphonsine Tables.*" He failed to indicate, however, where Copernicus could have found the length of the tropical year in his edition of the *Alfonsine Tables.* Had they been associated with $365^d5^h49^m$ in the mind of Copernicus, would he not have cited this famous work in the same breath with his three other illustrious authorities: Hipparchus, Ptolemy, al-Battani? If Copernicus had really had a choice between Alfonso X and Alfonso de Corduba Hispalensis, would he not have preferred the world-renowned royal patron of astronomy to the relatively obscure Hispalensis? In fact, however, Copernicus had no such choice. For, Hispalensis' length of the year was, as we saw above, "365¼ days minus 11 minutes." That subtrahend was not 11 minutes in the *Alfonsine Tables,* according to the author of the later *Prussian Tables,* but, "about 10 minutes 44 seconds, which is a little more than one-sixth of an hour" = 10 minutes.[31]

This astronomical divergence indicates Hispalensis' independence of the *Alfonsine Tables.* The latter work had no special connection with Seville (Hispalis). By contrast, Hispalensis' *Astronomical*

---

genannt *Regiomontanus,* 2nd ed., Osnabrück, 1968, p. 252, no. 7). Zinner (p. 251) misidentified the royal lady in Hispalensis' title with Queen Elizabeth I of England. Hispalensis, however, dedicated his *Astronomical Tables* to "Fernando et Elisabeth Hispanie et Sicilie Regibus."

[31] Erasmus Reinhold, ed. Peurbach's *Theoricae novae planetarum* (Wittenberg, 1542, sig. e4v-5r): "Annum enim faciunt [Alphonsini] 365 dierum cum quadrante minus 10 scrupulis 44 secundis fere, id quod paulo plus est sextante unius horae."

Tables for Queen Isabella "compute the mean motions for the meridian of Seville," present a "Table of the Rising of the Signs at Seville," and offer a star catalog for Seville and Rome (sig. *A1v, D8r-E2r, L4v). At sig. *A2v, the foundation of the city of Seville is attributed to Hercules' son, Hispalis.

In his effort to obtrude Alfonso X into the *Commentariolus,* Swerdlow belittled Hispalensis' *Almanach* of 1502, while ignoring his *Astronomical Tables* of 1503 as well as his *Lumen caeli sive expositio instrumenti astronomici a se excogitati* (Rome, 28 May 1498). The foregoing paleographical, historical, institutional, bibliographical, and stylistic considerations clash with Swerdlow's attempt to replace Hispalensis by Alfonso X in the *Commentariolus.* Neither in that work nor in the *Revolutions* did Copernicus ever mention Alfonso or the *Alfonsine Tables.*

Copernicus' lack of interest in the *Alfonsine Tables* may have been due to the emergence of competing tables. One such successful rival, the aforementioned *Prussian Tables,* computed after Copernicus' death and first published in 1551, was linked with Copernicus posthumously. Hence, when our present calendar was adopted in 1582, the *Alfonsine Tables* were preferred to the (supposedly Copernican) *Prussian Tables* "mainly because the *Prussian Tables* are based on hypotheses which are very uncertain, not to say irrational, and abhorrent to the general opinion of mankind and repudiated by all natural philosophers."[32] While the *Alfonsine Tables* remain at the basis of our calendar, there has been a shift in the general opinion of mankind regarding Copernicanism, which is no longer repudiated by all natural philosophers.

---

[32] Christopher Clavius, *Romani calendarii . . . explicatio,* Chapter 4, in Clavius' *Opera Mathematica* (Mainz, 1611-1612), V, 67.

5

Copernicus and Al-Bitruji

The twelfth-century Muslim scientist, Al-Bitruji, was mentioned only once by Nicholas Copernicus (1473–1543), the founder of modern astronomy. In his epoch-making treatise *De revolutionibus orbium coelestium* (Nuremberg, 1543) Copernicus demolished the traditional contrast between heaven and earth. Destroying the ancient misconception of the earth as differing in its essence from the celestial bodies, he recognized its true character as a satellite of the sun. Being therefore obliged to put it in its proper place within the cosmic order, he had to re-arrange the planets (*Revolutions*, Book I, Chapter 10). Seeking to make this re-arrangement more acceptable to his readers, he reached back into the history of the subject in order to show that famous men, who had agreed about the earth being the center of the universe, had disagreed about the position of Mercury and Venus. These two planets, he pointed out, had been located by Plato above the sun, and by Ptolemy below the sun. Then Copernicus continued: "Alpetragius superiorem sole Venerem facit, et inferiorem Mercurium" (Al-Bitruji makes Venus higher than the sun and Mercury lower).

Copernicus' remark about Al-Bitruji has been mistranslated as follows: "Alpetragius makes Venus nearer and Mercury further than the Sun." This transposition of Venus and Mercury in Al-Bitruji's planetary theory was initiated in 1947 by John Frederic Dobson (1875–1947), professor of Greek at the University of Bristol, and Selig Brodetsky (1888–1954), professor of applied mathematics at the University of Leeds[1].

Had the Dobson-Brodetsky transposition been printed only once, we could perhaps afford to disregard it, even though it appeared in the Royal Astronomical Society's Occasional Notes[2]. But it was reprinted unchanged when the Dobson-Brodetsky partial translation of Copernicus' *Revolutions*[3] was reissued by the Royal Astronomical Society in 1955. Two years later the Dobson-Brodetsky transposition was repeated twice: once by Milton K. Munitz, *Theories of the Universe*[4], and again by Thomas S. Kuhn, *The Copernican Revolution*[5]. In preparing his book for re-publication as a paperback, Kuhn explained that he had had time "to correct a number of errors", but he left the Dobson-Brodetsky transposition unaltered[6]. This has now come to my notice in print six times. Do we dare hope that its further spread can be checked?

The transposition was not the only error committed by Dobson-Brodetsky in connection with Copernicus' statement about Al-Bitruji. Recalling the *Book on the Sphere*, which the Muslim astronomer wrote in Arabic shortly after 1185, Dobson-Brodetsky blithely asserted that it was "this work, to which Copernicus here doubtless refers"[7]. But there is in fact the gravest doubt that Copernicus here refers to this work.

In the first place, Copernicus never learned Arabic. If he read Al-Bitruji at all, he could have done so only in Latin translation.

Al-Bitruji was translated into Latin twice[8]. The earlier of these two translations, done by Michael Scot in 1217, was printed for the first time just a few years ago[9]. Hence no printed copy of Michael Scot's translation of Al-Bitruji was available to Copernicus. Eleven manuscript copies are known, all of them written in the fourteenth and fifteenth centuries[10]. However, we should avoid the all too common mistake of supposing that just because a manuscript existed, it must have been read by somebody who happens to interest us. There is in fact no evidence that Copernicus ever laid eyes on any manuscript of Michael Scot's translation of Al-Bitruji.

The only other Latin translation was made by Calo Calonymos[11], not directly from the Arabic original, but from an intervening Hebrew version[12]. This second Latin translation of Al-Bitruji formed a separately paginated part of *Sphaerae tractatus*, a composite volume published at Venice in 1531.

The latest astronomical observation recorded by Copernicus in the *Revolutions* is dated 1529[13]. Although he made some later revisions, and although the time when he composed the various sections of the *Revolutions* cannot yet be determined with precision, there is every reason to believe that Book I, Chapter 10, with its reference to Al-Bitruji was finished long before 1531.

If the foregoing analysis is sound, Copernicus never saw Al-Bitruji's *Book on the Sphere*, neither the Arabic original, nor the Hebrew version, nor either of the two Latin translations. Despite Dobson-Brodetsky, it is not "this work, to which Copernicus here doubtless refers".

In that case, how did Copernicus find out about Al-Bitruji?

A useful clue is Copernicus' scanty acquaintance with the Arab astronomer. All he says about the Muslim who had once caused a sensation by climaxing the Islamic attack against Ptolemy in returning to the homocentric theory is a brief comment concerning Al-Bitruji's peculiar arrangement of Venus and Mercury.

"Al-Bitruji located Venus below Mars, the sun below Venus, and then Mercury," said the fifteenth century's greatest astronomer, John Regiomontanus (1436–1476), in his *Epitome of Ptolemy's Syntaxis*. This widely consulted and highly influential work was first published at Venice in 1496[14]. That was the year in which Copernicus arrived in Italy to continue his university studies[15]. Conclusive proof that Copernicus' reference to Al-Bitruji was based on Regiomontanus' *Epitome* was furnished long ago by Ludwik Antoni Birkenmajer (1855–1929)[16]. His understandable but regrettable decision to publish his masterly work in Polish, his native language, has had the lamentable effect of virtually sealing off his valuable researches from the mainstream of Copernican studies. Nevertheless, he translated into French his summary of his most important results; at the very head of his seventy-six theses, Birkenmajer put Copernicus' use of Regiomontanus' *Epitome* as the principal source of his information about Muslim astronomy in general and Al-Bitruji in particular[17].

Copernicus' limited knowledge of Al-Bitruji stands in striking contrast to Regiomontanus' much wider familiarity with the thinking of the Arab astronomer. Whereas Copernicus never saw a manuscript of Al-Bitruji, Regiomontanus had his own personal copy[18] of Michael Scot's translation. In this copy Regiomontanus wrote many comments, some of which have been published[19].

Moreover, in a separate manuscript, Regiomontanus composed a refutation of Al-Bitruji's astronomy[20]. In this refutation, which was published for the first time a decade ago[21], Regiomontanus mistakenly declared that Al-Bitruji placed Mercury above the sun[22]. This mistake, which was made by Regiomontanus in his youth, fortunately was confined to this single manuscript, which waited nearly 500 years to be printed.

Nor was the manuscript copied. Carmody was entirely wrong in saying[23] that Ernst Zinner left it doubtful whether Regiomontanus' refutation of Al-Bitruji was written in his own handwriting. His refutation forms part of a composite manuscript[24], which Zinner described as containing "others' works, copied by Regiomontanus and others, as well as essays and shorter computations"[25] by Regiomontanus himself. Zinner clearly classified the refutation of Al-Bitruji as one of Regiomontanus' "own works"[26].

Since Regiomontanus composed the refutation of Al-Bitruji and wrote it out with his own hand, and since it was never copied nor printed (until 1951), its mistake in saying that Al-Bitruji located Mercury above the sun occurred only once. This youthful mistake was not repeated in the *Epitome*, which Regiomontanus wrote later on and which was printed three times. In the *Epitome* Regiomontanus correctly stated that Al-Bitruji put Mercury below the sun[27].

A generous portion of his reaction to Al-Bitruji was incorporated by Regiomontanus in his *Epitome*. It was from that source, not from Al-Bitruji's *Book on the Sphere*, that Copernicus derived the meager remark which was turned topsy-turvy by Dobson-Brodetsky, Munitz, and Kuhn.

NOTES

1. Brodetsky, Selig: *Memoirs*, London: Weidenfeld and Nicolson, 1960, p. 105.
2. 1947: 2: 16.
3. *De revolutionibus, Preface and Book I.*
4. Glencoe, Illinois: Free Press, p. 165; London: Allen and Unwin, 1958.
5. Cambridge, Mass.: Harvard University Press, p. 176.
6. New York: Random House, 1959, pp. xii, 177.
7. R. Astr. Soc., Occas. Notes, 1947 (1955): 2: 30.
8. Carmody, Francis J.: *Arabic Astronomical and Astrological Sciences in Latin Translation*, Berkeley and Los Angeles: University of California Press, 1956, pp. 165–166.
9. *Al-Bitruji*, Latin tr. by Michael Scot, ed. F. J. Carmody, Berkeley and Los Angeles: University of California Press, 1952.
10. Carmody, in *Al-Bitruji*, pp. 160–161.
11. Roth, Cecil: *The Jews in the Renaissance*, Philadelphia: Jewish Publication Society of America, 1959, pp. 75–76, 160, 234.
12. *Alpetragii arabi planetarum theorica phisicis rationibus probata, nuperrime latinis litteris mandata a Calo Calonymos Hebraeo Neapolitano.* A microfilm of this translation was courteously procured for me by Professor Edward Grant, Department of History and Logic of Science, Indiana University, and was paid for by the Research Committee of the City College of New York.
13. Lunar occultation of Venus (*Revolutions*, V, 23).
14. *Epytoma Joannis de Monteregio in Almagestum Ptolomei*, ed. Venice, 1496, fol. klv: "Alpetragnis ... sub Marte Venerem, sub qua Solem, deinde Mercurium statuebat"; ed. Basel, 1543, p. 163; ed. Nuremberg, 1550, fol. M6v.
15. Prowe, Leopold: *Nicolaus Coppernicus*, Berlin, 1883–1884, vol. I, part 1, p. 230.
16. *Mikołaj Kopernik*, Kraków, 1900, p. 7.

17. Bulletin international de l'Académie des sciences de Cracovie, classes des sciences mathématiques et naturelles, 1902: 206–208. Rosen, Edward: *Three Copernican Treatises*, 2d ed., New York: Dover Publications, 1959, pp. 65, 67, cites examples of Copernicus' use of Regiomontanus' *Epitome*.

18. Nuremberg Municipal Library, Cent. V 53, fol. 74r–111v; Zinner, Ernst: *Leben und Wirken des Johannes Müller von Königsberg genannt Regiomontanus*, Munich, 1938, pp. 44, 219.

19. Zinner, *Regiomontanus*, pp. 44–45; Zinner: *Entstehung und Ausbreitung der copperni-canischen Lehre*, Erlangen, 1943, pp. 128–129.

20. Zinner, *Regiomontanus*, pp. 46–48, 220; *Entstehung*, p. 129.

21. Carmody: Regiomontanus' Notes on Al-Bitruji's Astronomy, Isis, 1951: 42: 121–130.

22. "Posuit deinde Alpetragnis ... Venerem et Mercurium supra Solem" (Isis, 1951: 42: 128, no. 29). Yet Michael Scot's translation, a copy of which was in Regiomontanus' possession, plainly said: "celi Mercurii et celi Lune ... sunt sub [celo Solis]. Sed videtur de re Veneris quod sit supra celum Solis et inter ipsum et inter celum Martis, licet antiqui posuerunt ipsum sub celo Solis" (*Al-Bitruji*, ed. Carmody, p. 128). The 1531 translation rendered this passage as follows: "orbis mercurii et orbis lunae existentium sub eo [sole]. Venus vero videtur supra orbem solis ac inter eum et martem, quamquam antiqui posuerunt eum sub orbe solis" (fol. 21r).

23. Isis, 1951: 42: 122.

24. Vienna, Austrian National Library, MS 5203.

25. *Regiomontanus*, p. 46.

26. Op. cit. (note 20), p. 47: "eigene Arbeiten."

27. See note 14, above.

6

Copernicus' Alleged Priesthood

INTRODUCTION

Nicholas Copernicus, the founder of modern astronomy, never called himself a priest throughout the seventy years of his life (1473–1543). Neither his friends nor his enemies, neither his close associates nor his distant acquaintances, ever referred to him as a priest. In short, the very idea of linking the astronomer with the priesthood had not even been dreamt of during the seven decades of his activity on earth.

An equal length of time elapsed after his death in 1543 before the misdescription of Copernicus as a priest was devised for the first time. The responsibility for this distortion of history rests squarely on the shoulders of Galileo Galilei (1564–1642). This great Italian scientist was trying hard to save the Roman Catholic church from making the grave mistake of censuring the Copernican astronomy. As a thinker, Galileo was convinced that Copernicus was right. As a Catholic, Galileo was devoted to the long-range interests of his church. In his intense effort to block the Catholic condemnation of Copernicanism, Galileo committed five serious historical mistakes concerning Copernicus.[1] Each of these five blunders had the effect of binding Copernicus more closely to the Catholic church than the facts themselves warranted.

Of particular interest to us is Galileo's misstatement that Copernicus was a priest. At first, however, instead of mislabeling Copernicus a secular priest, Galileo had previously assigned the astronomer to a religious order, that is, to the regular clergy. But this erroneous classification was promptly withdrawn by Galileo, who replaced it by the falsification "Copernicus the priest." Not the slightest shred of evidence was offered by Galileo to support either his original attempt to enroll Copernicus in the regular clergy or his later effort to ordain Copernicus a priest. No such evidence was adduced by Galileo because no such evidence ever existed. The falsification "Copernicus the priest" was invented by Galileo while he was embroiled in his bitter and tragic struggle with the Roman Catholic church. The falsification "Copernicus the priest" has nothing whatever to do with the historical Copernicus.

More than two centuries after Galileo's falsification and without any knowledge of it, a short story entitled "The Youth of Copernicus" was published in a Warsaw literary magazine by Anna Nakwaska (1781–1851).[2] In this frankly fictional piece, Nakwaska, a highly imaginative writer, drew a romantic picture of Copernicus as a young student on his way home from the university rescuing a fair princess in distress, who later visited him after he had become a priest.

Less than a decade later, in a letter in the 19 February 1843 issue of the Warsaw newspaper *Kurjer Warszawski*, Adrian Krzyżanowski (1788–1852), emeritus professor of the University of Warsaw, supplied far more detail than Nakwaska. Paying scant attention to the canons of historical scholarship, Krzyżanowski professed to know when, where, and by whom Copernicus was ordained a priest. Unfortunately for Krzyżanowski, the time and place he chose for the astronomer's pretended ordination turned out to be incompatible with each other. Krzyżanowski made the fatal mistake of having Copernicus ordained a priest in Cracow between 1502 and 1509. During those years, however, the astronomer was not in Cracow. Morever, of the two prelates selected by Krzyżanowski to preside over Copernicus' ordination, one never existed, being merely a patchwork produced by Krzyżanowski's unprincipled scrambling of ecclesiastical records.[3] Copernicus was not ordained a priest at Cracow between 1502 and 1509, as Krzyżanowski tried to hoodwink his readers into thinking.

The third and, let us hope, the last such deception was perpetrated by the Italian historian Lino Sighinolfi (1876–1956). Having discovered a notarial document in which the notary declares that Copernicus "appeared in person" (*personaliter constitutus*),

---

1. Edward Rosen: "Galileo's Misstatements about Copernicus," *Isis*, XLIX (1958), 319–330.; below, 193–204.

2. *Jutrzenka*, 1834, pp. 209–255.

3. Edward Rosen: "Copernicus Was Not a Priest," *Proceedings of the American Philosophical Society*, CIV (1960), 636–640.

Sighinolfi made believe that the document described Copernicus as "having been made a priest" (*presbiter constitutus*).[4] After Sighinolfi's fraudulent substitution of *presbiter* for *personaliter* has been rectified, his document, which is based on information supplied to the notary by Copernicus himself, correctly classifies the astronomer as a canon, but not as a priest.

A little more than five years later, on 10 January 1503, Copernicus presented himself officially as a "canon of Varmia [Ermland] and scholaster of the church of the Holy Cross in Breslau," but not as a priest.[5] Finally, in a document dated 30 March 1519, a colleague of Copernicus empowered thirteen proxies to act for him, calling some of them priests and giving the remainder other ecclesiastical titles; among these non-priests was Copernicus.[6] The half-millennium since his birth in 1473 provides absolutely no evidence that the astronomer was a priest, but it does show us three distinct falsifications of history designed to deceive unwary readers into believing that Copernicus was a priest.

I. The Thirteen Proxies of Alexander Scultetus (Schultze)

The aforementioned document dated 30 March 1519 was declared by the late Hans Schmauch (1887–1966), honorary professor of East German Church and National History at the University of Mainz, to be "without importance" (*belanglos*) for our investigation.[7] Schmauch reached this startling conclusion by deleting from this highly important document[8] both its uses of the word "priest."

In its first occurrence the word appears as *presbiterum*. Schmauch admitted that this was "unequivocally the paleographical reading" (*im Urkundentext einwandfrei*, p. 426). Nevertheless, he decided that this paleographically unobjectionable *presbiterum* was "probably a scribal error" (*wahrscheinlich verschrieben*, p. 430).

In its second occurrence the word appears in the plural form *presbiteros*. This time Schmauch somehow could not think up any suitable scribal error. Instead, he resorted to a different strategy. Imagining that he had already successfully eliminated *presbiterum*, he remarked that thereafter "the designation 'priest' is applied solely to the extra-Varmian ecclesiastics, who are omitted from consideration here."[9]

Schmauch's procedure is memorable as a shining example of how not to write history. In the discussion of Copernicus' thrice-mythical priesthood a document had been introduced because it mentions the astronomer as a non-priest by contrast with other eccle-

---

4. Lino Sighinolfi: "Domenico Maria Novara e Nicolò Copernico allo Studio di Bologna," *Studi e memorie per la storia dell' Università di Bologna*, V (1920), 205–236.

5. Erice Rigoni: "Un autografo di Niccolò Copernico," *Archivio veneto*, XLVIII to XLIX (1951), 147–150.

6. Guido Horn-d'Arturo: "Atti notarili del sec. XVI contenenti il nome di Copernico rinvenuti nell' Archivio storico capitolino," *Coelum*, XIX (1951), 40–43.

7. Hans Schmauch: "Um Nikolaus Coppernicus," pp. 417–431 in *Studien zur Geschichte des Preussenlandes, Festschrift für Erich Keyser* (Marburg: N. G. Elwert Verlag, 1963), p. 427.

8. Reprinted by Schmauch, pp. 430–431.

9. Schmauch, p. 427: "Die Kennzeichnung *presbiter* ... lediglich bei den hier ausser Betracht bleibenden fremden Geistlichen verwendet ist."

siastics who are priests. Schmauch then sought to deprive this document of its great importance by emending *presbiterum* and closing his eyes to *presbiteros*. Omitting crucial evidence from consideration may be beneficial to the eyes but is surely harmful to serious historical research.

Let us now take a closer look at the crucial evidence in the document dated 30 March 1519. On that day in Rome Alexander Scultetus (Schultze) named thirteen proxies to help him obtain possession of "the canonry and prebend in the church of Varmia which were held by the late Andrew Copernicus while he was alive."[10] The astronomer Nicholas Copernicus, younger brother of Andrew Copernicus, was naturally included by Scultetus among his thirteen proxies.

Of these thirteen proxies, Scultetus chose nine from the church of Varmia, and the other four from the dioceses of Leslau and Mainz. As a group, these last four proxies were described as "priests and clerics" (*presbiteros et clericos*). By contrast none of the Varmian ecclesiastics was described as a cleric, and only one as a priest (*presbiterum*). Of the remaining eight Varmian proxies, three were vicars, and five were canons, including Nicholas Copernicus.

For the purpose of positive identification the thirteen proxies were described by Scultetus as belonging to a specific diocese and as possessing a specific ecclesiastical status. The four "priests and clerics" from Leslau and Mainz, however, were lumped together. In their case, by contrast with the nine proxies from Varmia, the diocese and status were indicated for the group of four extra-Varmian proxies as a whole, not for each individual separately. Under these circumstances let us assume that Scultetus chose two proxies from Leslau and two from Mainz, each pair consisting of a priest and a cleric.[11] If our assumption is correct, and it surely cannot be far wrong, two of the extra-Varmian ecclesiastics were priests. These proxies, whatever their exact number may have been, were described by Scultetus as priests, whereas he called Copernicus a canon, not a priest. Like Copernicus himself, like every other contemporary who knew our astronomer, Scultetus labeled him a canon, not a priest.

Scultetus' crucial evidence lifts his proxy document of 30 March 1519 to the level of highest importance for our investigation. By omitting Scultetus' extra-Varmian priests from consideration, Schmauch violated the most fundamental rule of historical research: consider all relevant evidence, do not disregard any of it. Schmauch claimed that Scultetus' extra-Varmian proxies could be properly excluded (*können ... füglich ausser Betracht bleiben*) from his investigation, which was concerned exclusively with Varmia (*sich ja nur auf das Ermland bezieht*, p. 426). But Scultetus' document was not concerned exclusively with Varmia, since it named four extra-Varmian proxies, some of whom were priests.

Having examined Schmauch's harmful neglect of the extra-Varmian priests (*presbiteros*), let us next consider his proposed emendation of *presbiterum* in the Varmian context. At the head of his nine Varmian proxies, Scultetus put the name of Christopher von Suchten, whom he described as a priest (*presbiterum*). This manuscript reading was

---

10. *Coelum*, p. 41; Schmauch, p. 431.

11. Jerzy Sikorski: *Mikołaj Kopernik na Warmii* (Olsztyn, Stacja naukowa polskiego towarzystwa historycznego, 1968), pp. 152–153, assigned all four of these extra-Varmian proxies to Leslau (Włocławek), making two of them priests and the other two clerics. If so, who were Scultetus' proxies in Mainz?

acknowledged by Schmauch, as we have already seen, to be unobjectionable (*einwand-frei*, p. 426). However, Schmauch objected to *presbiterum* for the following reason. Of the eight Varmian proxies other than Suchten, five were canons and three were vicars. Arguing that "canon" and "vicar" each "designated an office" (*Amtsbezeichnung*, p. 427), Schmauch insisted that like the other eight Varmian proxies Suchten too must have a "designation of an office." Since "priest" did not designate any office in the Chapter, Schmauch contended that *presbiterum* was "probably a scribal error for provost,"[12] the office held by Suchten in the Varmian Chapter from 1513 to 1519.

In so arguing, Schmauch overlooked the motive of Scultetus' descriptions of his proxies. Scultetus was primarily concerned with the unambiguous identification of each of these persons. For this reason he called Suchten a priest, a title seldom held by members of the Varmian Chapter and therefore very useful for purposes of identification. The title "priest" was similarly applied by Scultetus, let us recall, to some of his extra-Varmian proxies. When we consider the whole of Scultetus' document, instead of merely a part of it in Schmauch's self-blinding manner, we see clearly that Scultetus called one of his Varmian proxies a priest, and the other eight Varmian proxies canons and vicars, that is, non-priests. One of these non-priests was Copernicus.

To support his suggestion that in Scultetus' document *presbiterum* was a scribal error, Schmauch remarked that the document exhibits "various scribal errors, especially in the proper names."[13] Actually in this document of considerable length Schmauch found the grand total of three errors: two surnames (Libernam and Ficihl, for which he suggested Libenaw and Reich), and "i" as the third letter in the name of the diocese *Wladis-laviensis* (Leslau).

Although the scribe was admittedly not letter-perfect, on the whole he copied Scultetus' document rather well. True, he found three unfamiliar German and Polish proper names somewhat difficult. But such stumbling over exotic nomenclature does not justify Schmauch's supposition that the scribe had trouble with the ordinary Latin words *presbiterum* and *prepositum*. In the case of these two words Schmauch suggested that the scribe committed "an error in reading or writing" (*einen Lese- oder Schreib-fehler*, p. 427).

In opposition to Schmauch, I would suggest that the scribe committed no such error, because Scultetus actually called Suchten a priest (*presbiterum*). Failing to understand Scultetus' perfectly valid reason for doing so, Schmauch sought to emend *presbiterum* out of Scultetus' document. In like manner Schmauch closed his eyes to Scultetus' extra-Varmian priests (*presbiteros*). For those who prefer to keep their eyes open, however, the extra-Varmian priests are present in Scultetus' document. Since it similarly describes Suchten as a priest, we must discard Schmauch's proposed elimination of *presbiterum* as paleographically unjustified and diplomatically unsound. At the same time we must reject Schmauch's evaluation of Scultetus' document as "without importance" for the investigation of Copernicus' thrice-mythical priesthood. On the contrary, Scultetus' document is highly important precisely because it calls some of Scultetus' proxies priests and Copernicus a non-priest.

12. Schmauch, p. 430: "Wahrscheinlich verschrieben statt *prepositum*."
13. Schmauch: p. 427: "namentlich bei den Eigennamen mancherlei Verschreibungen."

II. Schmauch's Attitude Toward the Myth of Copernicus' Priesthood

"Even without this [Scultetus] document of 1519 there really is no doubt that the astronomer did not receive ordination as a priest."[14] This was in 1963 Schmauch's final conclusion, the correctness of which he had previously failed to see.

Nevertheless, Schmauch did make a valuable contribution to unmasking Sighinolfi. At the Staatliche Akademie zu Braunsberg, where Schmauch taught from 1932 to 1945, his colleague Herman Hefele (1885–1936)[15] uttered doubts (*Bedenken*) about Sighinolfi's reading *presbiter constitutus*, "but only in the course of conversation" (*freilich nur gesprächsweise*, p. 421). Although familiar for years with these orally expressed misgivings about Sighinolfi's *presbiter constitutus*, Schmauch had Copernicus ordained a priest three times in print.[16] Then in June 1942 Schmauch personally inspected Sighinolfi's document and corrected *presbiter* to *personaliter* in a 1943 collective work.[17]

This product of the period when our two countries were at war with each other was not available to me when I published my article in 1960. I sincerely regret that I overlooked this article by Schmauch while I cited eight other articles by him as well as a book, both in the original German and in English translation. The passage overlooked by me, as Schmauch himself generously acknowledged, appeared "only in a footnote" as "a very brief reference" "really quite buried."[18] Schmauch called his brief footnote "buried" because in this collective work his article was printed at pages 233–256, but the footnote on page 370 as note 8.

Although this inconspicuous footnote of 1943 was overlooked by me, in 1960 I quoted from Schmauch's book of 1953 his admission that "there is no definite proof that Copernicus ever took holy orders."[19] Asking myself at that time why Schmauch had abandoned his previous priestly ordination of Copernicus, I suggested that "perhaps Schmauch recognized the Sighinolfi ... blunder when he saw the words *personaliter ... constitutus* in a document executed by the notary" who also drew up the legal paper distorted by Sighinolfi. Without being aware that in fact Schmauch had replaced Sighinolfi's *presbiter* by *personaliter*, I assumed that Schmauch had noticed Sighinolfi's misrepresentation.[20]

---

14. Schmauch, p. 428: "Auch ohne dies Dokument von 1519 besteht eigentlich kein Zweifel, dass der Astronom nicht den priesterlichen Weihegrad besessen hat.'

15. Hermann Binder: "Herman Hefele," *Dichtung und Volkstum*, XXXVIII (1937), 157–171.

16. To the two passages recalled by Schmauch (p. 421, n. 23) may be added *Jomsburg*, I (1937), 174.

17. Fritz Kubach, ed.: *Nikolaus Kopernikus, Bildnis eines grossen Deutschen* (Munich and Berlin, 1943), containing Schmauch's article "Nikolaus Kopernikus und der deutsche Osten."

18. Schmauch, p. 422: "nur in einer Anmerkung ... ein ganz kurzer Hinweis ... freilich recht versteckten Hinweis."

19. Schmauch: *Nikolaus Kopernikus* (Kitzingen, 1953; Der Göttinger Arbeitskreis, Schriftenreihe Heft 34), p. 15: "Es fehlt übrigens jeder Beweis dafür, daß Kopernikus die Priesterweihe empfangen hat."

20. Rosen: "Copernicus Was Not a Priest," p. 657.

"A bad misreading" (*ein böser Lesefehler*) is what Sighinolfi was charged with in Schmauch's 1943 footnote. Even in 1963 Schmauch still implied that Sighinolfi did not know how "to expand correctly the abbreviations found" in documents.[21] Sighinolfi was in fact a highly experienced archivist, however, and knew perfectly well the difference between "ptr" (= *presbiter*) and "psolr" or "psonalr" (= *personaliter*). The deliberateness of Sighinolfi's substitution of *presbiter* for *personaliter* was not recognized by Schmauch, presumably because he neglected to give adequate weight to Galileo's false assertion of Copernicus' priesthood and was likewise insufficiently informed about Krzyżanowski's more elaborate falsification. In other words, Schmauch failed to recognize in Sighinolfi the third wave in the recurrent campaign to distort history by conferring the priesthood on Copernicus.

In 1943 Schmauch corrected Sighinolfi's *presbiter* to *personaliter*. How did Schmauch then proceed to use his correction? Ten years later, in his aforementioned book, Schmauch said, as we saw above, that "there is no definite proof that Copernicus ever took holy orders." Schmauch repeated this statement in *Neue Deutsche Biographie*, III, 349.[22] In 1953 and 1957, therefore, Schmauch used his 1943 correction merely to remark that proof of Copernicus' priesthood was lacking.

It is one thing to say that proof of a historical statement is lacking: the statement may still be true, history being far from a mathematical science. It is quite another thing to say that a historical assertion is false. Calling attention to an assertion's lack of proof is by no means equivalent to an outright denial of the assertion. It was only in 1963, after Schmauch had learned of my 1960 article, that he denied Copernicus' priesthood.[23]

Then, having declared that there is no doubt (*kein Zweifel*) about the matter, five lines later Schmauch reintroduced doubt by saying that "there is no proof that the astronomer ever received ordination as a priest."[24] This conclusion was attributed by Schmauch to Franz Hipler writing in 1868. But Hipler did not make the statement attributed to him by Schmauch. That statement was first made by Adolf Müller in 1898.[25] In 1868 Hipler said: "in the case of Copernicus ... it is questionable whether he ever took the three major orders," deacon, priest, bishop.[26]

In like manner, according to Schmauch, Scultetus' proxy document was discovered by Guido Horn-d'Arturo.[27] But Horn-d'Arturo himself gave the credit for the discovery of Scultetus' document to someone else.[28]

---

21. Schmauch, p. 421: "die vorkommenden Abkürzungen richtig aufzulösen."

22. This volume was published in 1957, not 1937, the year mistakenly given by Schmauch (1963, p. 422, n. 25).

23. See n. 14, above.

24. Schmauch, p. 428: "es gibt keinen Beweis dafür, dass der Astronom je die Priesterweihe erhalten hat."

25. Rosen: "Copernicus Was Not a Priest," p. 647.

26. Hipler: "Nikolaus Kopernikus und Martin Luther," *Zeitschrift für die Geschichte und Alterthumskunde Ermlands*, cited hereafter as ZGAE, IV (1867–1869), 502: "bei Kopernikus ... ist es fraglich, ob er die drei höhern [Weihen] jemals empfangen habe."

27. Schmauch, p. 426: "eine bisher nicht bekannte Urkunde, die durch den italienischen Gelehrten Guido Horn-d'Arturo im Archivio Capitolino in Rom neu aufgefunden ... worden ist."

Scultetus' description of Christopher von Suchten as a priest was repeated by Horn-d'Arturo in reliance on Schmauch.[29] But when the City Council of Danzig on 5 September 1509 asked Pope Julius II to permit control of the parish church of St. John to pass into the hands of Suchten, he was not described as a priest. He enjoyed the income from the parish *(genoss die Pfarreinkünfte)* from 1509 until 1516 without setting foot in Danzig *(wird nie in Danzig genannt).*[30]

Suchten was in Rome on 27 August 1510 when as a Varmian canon he wrote to his bishop Fabian von Lossainen, commending himself and his entire family *(me meamque familiam omnem).*[31] On 23 January 1511 a papal bull issued by Julius II permitted Suchten to possess benefices in "cathedral, metropolitan, collegiate, parish, and other churches, even if any of the benefices should be connected with the care of souls. ... When he has become a priest, he shall be authorized to absolve ... [certain] sins. He may reside in Rome and elsewhere for purposes of study, without being bound by the residential obligation connected with the benefices."[32] On 23 January 1511 Suchten was in Rome, in the service of a dignitary highly placed in the papal court. Presumably the papal bull of 23 January 1511 was drafted with the full knowledge of Suchten, and perhaps with his cooperation. It clearly implies that he was not a priest on 23 January 1511.

Later that year, in October or November, when Suchten was admitted to the College of Notarial Scribes of the Archive of the Roman Curia *(Collegium scriptorum archivii Romanae curiae notariorum),* he described himself as a "cleric of the Leslau diocese," not as a priest.[33] By contrast, two of Suchten's colleagues did describe themselves as priests *(presb. Basilien.,* November 1510; *presb. Tornacen. dioc.,* 1516).[34]

Suchten's status as a cleric gave him all the qualification he needed to become provost of the Varmian Chapter, the post conferred on him by Pope Leo X on 2 December 1513.[35] His predecessor had been provost nearly three years "without canonical right ... since he pretended to be a cleric."[36] In order to be provost of the Varmian Chapter, Suchten had to be a cleric, and indeed he was. But in order to be provost, Suchten did not have to be a priest, and in fact he was not.

In that case, it may be asked, why did Scultetus describe Suchten as a priest on 30 March 1519? Some years before, Suchten had left Rome. In 1515, he was in Breslau

---

28. Carlo Grigioni, *Coelum,* p. 40.

29. *Coelum,* p. 42, n. 14.

30. Theodor Hirsch: *Die Ober-Pfarrkirche von St. Marien in Danzig* (Danzig, 1843), I, 245.

31. Joseph Kolberg: "Der ermländische Dompropst Christoph von Suchten († 1519)," *Römische Quartalschrift für christliche Altertumskunde,* Supplementheft XX (1913), 166.

32. Kolberg, pp. 153–154.

33. Karl Heinrich Schäfer: "Deutsche Notare in Rom am Ausgang des Mittelalters," *Historisches Jahrbuch,* XXXIII (1912), 730.

34. Schäfer, pp. 730, 734.

35. Joseph Hergenroether: *Leonis X. pontificis maximi regesta* (Freiburg im Breisgau, 1884–1891), I, 350, no. 5587.

36. Kolberg, p. 156.

on 5 October,[37] and on 16 November in Frauenburg,[38] where he took up residence. In Rome on 30 March 1519 Scultetus was not accurately informed about Suchten's status. Perhaps Scultetus knew that on 8 September 1513 Pope Leo X had confirmed the grant of the parish church of Saints Lawrence and Elizabeth in Breslau to Suchten.[39] From this papal confirmation Scultetus may have drawn the erroneous inference that Suchten was a priest.

Such appointees "paid practically no attention to their parishes. ... These posts were regarded as a means of providing for the sons of the leading families, who in addition tried to acquire as many other benefices as possible. They spent the income from their benefices in some pleasant place at their ease, devoted either to purely sensual pleasure or to finer intellectual delights arising from occupation with science and art. They allowed their posts to be attended by poorly paid substitutes, who naturally accepted their parishes not exactly with enthusiasm and were frequently little respected. ... From the very beginning Suchten had no intention to devote himself to his priestly office."[40]

This realistic picture of Suchten's attitude toward his various parishes – multiple sources of income, not ecclesiastical obligations – is spoiled by the unhistorical assumption that Suchten was a priest, which was shared by Scultetus. In Rome on 30 March 1519 Scultetus did not foresee that five days later in Frauenburg Provost Suchten would notify the Varmian Chapter that he had to travel in order to seek medical advice.[41] Although Scultetus on 30 March 1519 could not have foreseen that Suchten would die within the next four months,[42] Scultetus did take the sensible precaution of naming twelve other proxies. By contrast with Suchten, whom Scultetus (mistakenly) described as a priest, most of the other twelve proxies were non-priests.

Thus we see that Scultetus' proxy document of 30 March 1519 is extremely important for our investigation. Schmauch's conclusion that this document was unimportant rested on two distortions of it. Like Schmauch, Scultetus (mistakenly) called Suchten a priest. Schmauch's effort to eliminate this occurrence of "priest" from Scultetus' document is paleographically unsound, as Schmauch himself admitted, and diplomatically unjustified. Schmauch's further neglect of Scultetus' extra-Varmian priests is historically unpardonable. Scultetus' document, in agreement with all our other evidence, decisively proves that Copernicus was not a priest.

---

37. Kolberg, p. 169.

38. Schmauch: "Das Präsentationsrecht des Polenkönigs für die Frauenburger Dompropstei," ZGAE, XXVI (1936–1938), 100.

39. Hergenroether, I, 268, no. 4403.

40. Paul Simson: *Geschichte der Stadt Danzig* (Danzig, 1913–1924; reprinted, Scientia Verlag, Aalen, 1967), I, 378–379.

41. Anton Eichhorn: "Die Prälaten des ermländischen Domcapitels," ZGAE, III (1864–1866), 319.

42. Suchten's death is mentioned in letters written by Grand Master Albert of the Teutonic Knights to Bishop Fabian von Lossainen and Archdeacon Johannes Scultetus on 26 July 1519; see Schmauch, ZGAE, XXVI (1936–1938), 102.

7

Copernicus was not a 'Happy Notary'
(with the Assistance of Erna Hilfstein)

A LETTER WRITTEN by Nicholas Copernicus (1473–1543), the founder of modern astronomy, was taken in 1820 from Poland to Great Britain by Karol Sienkiewicz (1793–1860). In his *Dziennik podróży po Anglii 1820–1821* (Diary of a Trip through England 1820–1821; not published until 1953), Sienkiewicz reports that he brought Copernicus' letter to Edinburgh. The *Edinburgh Philosophical Journal* promptly published a facsimile and translation of the letter from Latin into English. Although the translator was not identified, presumably he was Sienkiewicz's acquaintance, William Day, about whom little is known. Where Copernicus' letter says that

> certain business and compelling reasons require us, Felix and me, to remain at our stations *(d. felicem et me negotia quaedam et causae necessariae nos cogant in loco manere)*

the translator misinterpreted *d[ominum] felicem* as "my excellent protector."[1] A century and a half ago the history of the Cathedral chapter to which Copernicus and his friend Felix Reich belonged as fellow-canons was so little known that the translator may be readily forgiven for failing to recognize that Felix was a canon's baptismal name, and not an adjective. Contributing to the translator's pardonable blunder was Copernicus' use of a lower-case "*f*" as the initial letter of his fellow-canon's given name, a common practice when this letter was written on April 11, 1533.

In compliance with papal policy the Cathedral chapter of Varmia (Ermland), of which Copernicus and Felix Reich were canons, preserved important documents affecting its status and rights. As these documents increased in number, the chapter deemed it desirable to compile an inventory of them. Since Frombork (Frauenburg), the site of the cathedral, was situated close to the border between Varmia and its hostile neighbor, the Order of Teutonic Knights, as a precautionary measure the documents were moved south in 1502 to Olsztyn (Allenstein), the site of the chapter's most heavily fortified castle. The wisdom of this transfer was later confirmed in the war of 1520–1521, when Frombork was sacked and Olsztyn remained intact.

[1] *Edinburgh Philosophical Journal*, V (1821), 64.

In Olsztyn a revised inventory (the fourth in the history of the chapter) was prepared by its administrator in 1508. This 1508 inventory contains the following entry:

Copia una in pergameno de pace inter regem Ladislaum Poloniae etc. et ordinem facta anno MCCCCXI (One copy on parchment of the peace [treaty] between King Ladislaus of Poland etc. and the Order [of Teutonic Knights] that was made in the year 1411)

In the blank space between this and the next entry the following note was added later:

Haec copia missa fuit domino episcopo per d[ominum] Nic[olaum] Coppernic felicem notarium de voluntate dominorum visitatorum anno XI (This copy was sent to the bishop by Nicholas Copernicus [and] Felix [Reich], the notary, in accordance with the desire of the inspectors in the year [15]11)

The writer of this note was recently misidentified by Bishop Jan Obłąk with Copernicus.[2] His handwriting is readily accessible in *Nicholas Copernicus Complete Works*, Volume I (London/Warsaw, 1972), which provides a facsimile of Copernicus' autograph manuscript of his *De revolutionibus orbium caelestium*. After comparing Copernicus' handwriting in the 212 folios of his autograph manuscript with the handwriting in the two lines of this note, any unprejudiced scholar will surely conclude that someone other than Copernicus wrote the note.

That other someone was Tiedemann Giese (1480–1550), Copernicus' lifelong friend and fellow-canon, who served the chapter as its administrator from 1510 to 1515. Rather than prepare an entirely new inventory to supersede the 1508 inventory which he had inherited from the preceding administrator, Giese chose instead to add notes to the 1508 inventory. Among the notes so added is the one quoted above. The handwriting in this note may be readily compared with Giese's handwriting as reproduced in facsimile in two recent publications.[3]

The 1508 entry, it will be recalled, concerns a copy of a peace treaty, which "Copernicus copied with his own hand," according to Bishop Obłąk.[4] But Copernicus is not known to have set foot in Olsztyn before 1511, that is, three years after the copy of the peace treaty was registered in the inventory of 1508. The presence of this entry in the 1508 inventory of course indicates that the copy had been made before 1508.

---

[2] J. Obłąk, "M. Kopernika inwentarz dokumentów w skarbcu na zamku w Olsztynie" (N. Copernicus' Inventory of Documents in the Treasury of Olsztyn Castle), *Studia Warmińskie*, IX (1972), 15–16.

[3] Marian Biskup, ed., N. Copernici *Locationes mansorum desertorum* (Olsztyn, 1970), Plate XV; M. Biskup, ed., *Nowe materiały do działalności publicznej M. Kopernika* (Warsaw, 1971), Plate XXIVe.

[4] P. 15.

Indeed, because of its great historic importance to the chapter, this copy of the 1411 peace treaty may well have been among the documents recorded in the chapter's earliest inventory, which is dated about the middle of the fifteenth century.

Copernicus neither copied the peace treaty nor wrote the note concerning it. Yet Bishop Obłąk's twin misattributions of the copy and the note to Copernicus are far less disquieting than his misinterpretation of the note which, according to him (p. 15), states that the copy of the peace treaty was sent by Copernicus, the happy notary. Throughout his long career Copernicus never qualified as a notary. Like the translator in the *Edinburgh Philosophical Journal,* Bishop Obłąk converts Reich's baptismal name Felix into an adjective, thereby making him disappear from the note. But on December 28, 1512, Reich described himself as a "public notary by the sacred authority of the pope and the emperor" *(publicus sacris apostolica et imperiali auctoritatibus notarius).*[5] While Reich is banished from Bishop Obłąk's version of the note, a notary still remains. In that role Bishop Obłąk casts Copernicus (who lacks the necessary qualifications). With the disappearance of Reich from the note, and the transformation of his baptismal name into an adjective, Copernicus emerges in Bishop Obłąk's version (p. 16) as a happy notary *(szczęśliwy notariusz).*

Why was Copernicus, the supposed notary, happy? In Bishop Obłąk's fertile imagination Copernicus the notary was happy because he had the opportunity to handle a document that brought him into sympathetic rapport with King Ladislaus of Poland, who crushed the Teutonic Knights in the battle of Tannenberg/Grunwald in 1410, and then made peace with them in 1411.

But the happy notary's euphoria did not endure: the two lines stating that the peace treaty was sent to Copernicus' bishop were deleted. Who was responsible for this deletion? Copernicus, according to Bishop Obłąk. Why did he delete these two lines? Copernicus, forsooth, believed it to be conduct unbecoming a notary to show his feelings in official documents, because a well-behaved notary must preserve his impartiality at all times.[6]

Copernicus' joy and his withdrawal symptoms are of course purely fictitious. In sober reality the chapter possessed only one copy of this very important peace treaty.[7] When the two *visitatores* were in Olsztyn at the beginning of 1511, they expressed the wish that the

[5]Biskup, *Nowe materiały,* p. 34; cf. Harry Bresslau, *Handbuch der Urkundenlehre für Deutschland und Italien,* 4. ed. (Berlin, 1969), I, 628.

[6]Obłąk, p. 16, n. 21.

[7]According to M. Biskup, ed., *Regesta Copernicana, Calendar of Copernicus' Papers (Studia Copernicana,* VIII; Wrocław, 1973), p. 219, no. 64a, "Copernicus . . . delivers a duplicate of a copy of the peace treaty." But there was only one copy, and no duplicate.

chapter's copy of the treaty should be sent to the bishop in Lidzbark (Heilsberg) by Copernicus the *visitator* and Reich the notary. Then after the copy had been returned from Lidzbark to Olsztyn, as administrator from 1510 to 1515, Giese noted that the document was still there by writing *est* in the right margin alongside the entry. It was still in Olsztyn when Copernicus as administrator wrote his own inventory in 1520. His description of the document is a little shorter than the 1508 entry: he omits *una, in pergameno, facta anno MCCCCXI,* and he uses the symbol + instead of *ordinem.* Nevertheless, the document recorded in the 1508 inventory is unquestionably identical with Copernicus' *Copia de pace inter Vladislaum regem Poloniae et* +.[8] In fact, the document is preserved to this very day in the Olsztyn diocesan archives under the designation V8. When Copernicus handled it, he exhibited no special joy. He simply filed it in drawer K of the chapter's chest of drawers.

In Bishop Obłąk's faulty version of the note, not only does Felix Reich vanish but the *visitatores* suffer a like fate. They disappear when Bishop Obłąk (p. 15) misconstrues *de voluntate dominorum visitatorum* as "with the consent of the chapter" *(za zgodą Kapituły),* thereby confusing the part with the whole. For the entire chapter consisted of 16 canons, while at any one time it had only two *visitatores.* They were canons chosen by the chapter to carry out the annual task of inspection at the beginning of every year. It so happens that in 1511, the year mentioned in our note, Copernicus was one of the chapter's two *visitatores:*

> In the year 1511 by order of the venerable chapter we, Fabian of Lussigein and Nicholas Copernicus, assigned as inspectors *(visitatores)* by the venerable chapter, in Olsztyn at the celebration of the Lord's circumcision [January 1], received the remaining money on deposit in the castle for the vicariates of the venerable Zacharias [Tapiau], namely, 238 ¾ marks. And by order of the chapter we handed this money to the venerable B[althasar] Stockfyss on our return to the cathedral [in Frombork].[9]

The plan to send the copy of the peace treaty from Olsztyn to the bishop in Lidzbark sprang from the desire of the two *visitatores,* Fabian of Lossainen (Lussigein, Lutzingheim) and Copernicus, who were present in Olsztyn on temporary assignment, not from the consent of

[8]Obłąk, p. 47.

[9]Leopold Prowe, *Nicolaus Coppernicus* (Osnabrück, 1967; reprint of the 1883–1884 edition), published this extract in two slightly different versions: I[2], 256, n. **, omits Zacharias' name, and gives the fraction as 3½/4, by contrast with I[1], 381, n. **. Jerzy Sikorski, *M. Kopernik na Warmii* (Olsztyn, 1968), p. 28, no. 55, misprints *Venerabilis* and *pecuniam.* For Tapiau's establishment of the vicariates, see *Zeitschrift für die Geschichte und Altertumskunde Ermlands,* XIX (1916), 817–818.

the chapter in Frombork.

The harm done to the serious scholarly study of Copernicus' career by Bishop Obłąk's aberrations was magnified by the enthusiastic acceptance of "Copernicus the happy notary" (Mikołaj Kopernik, *felix notarius*) as a "sensational discovery" *(Rewelacyjne Odkrycie)*.[10] The present article was written in the hope that it might help to check the further spread of such uncritical enthusiasm.

[10]Piotr Bańkowski, in *Archeion*, IX (1974), 61-80.

8

Copernicus' Attitude toward the Common People

With a contemptuous side-glance at the masses Copernicus said that he wrote only for mathematicians. Very well known is Copernicus' statement that the masses mock him because they do not desire what he desires, and he does not desire what the masses desire. Rheticus had a similar attitude. He expressed it on the title page of Copernicus' work with the Platonic pronouncement: "Without a knowledge of geometry do not enter here." For Rheticus too, science for the sake of science stood in the foreground, apart from its uses.[1]

So we are told in a recent biography of George Joachim Rheticus (1514–74), the first disciple of Nicholas Copernicus (1473–1543), the founder of modern astronomy.

Did Copernicus cultivate science for the sake of science, apart from its uses? On July 27, 1501 he appeared in person at a meeting of the Cathedral Chapter of Frombork (Frauenburg), of which he was a member canon. Having completed three years of study in Italy with his Chapter's permission, he was now asking for two additional years. This extension was granted to Copernicus by his fellow canons "principally because Nicholas promised to study medicine, and as a helpful physician would some day advise our most reverend bishop and also the members of the Chapter."[2] Copernicus did not promise to study medicine for the sake of medicine, or for the sake of advancing the frontiers of medical knowledge. On the contrary, he looked forward to using his medical skill in the treatment of his fellow ecclesiastics when they were ill. And in fact he did so on numerous occasions after his return from studying medicine in Padua. On the other hand, he never conducted any medical research. We shall have occasion to take another look at his medical practice later on.

Although Copernicus made no contribution to the advancement of medi-

---

[1]Karl Heinz Burmeister, *G. J. Rhetikus* (Wiesbaden, 1967–68), I, 29–30. The presence of the Platonic pronouncement on the title page, it will be observed, is here attributed to Rheticus, the first editor of Copernicus' *Revolutions*. However, Rheticus left Nuremberg, where the *Revolutions* was being printed, for Leipzig University, which had just appointed him professor of advanced mathematics, that is, astronomy. In order to be present at the beginning of the winter semester, Rheticus arrived in Leipzig in Oct. 1542 (*Acta rectorum universitatis studii Lipsiensis 1524–1559*, ed. Zarncke [Leipzig, 1859], p. 196, line 24). Hence the front matter of the *Revolutions*, including the title page, was printed after his departure from Nuremberg. Thus he first saw the interpolated preface, to which he objected so strenuously, when complete copies of the *Revolutions* reached him in Leipzig in the spring of 1543. Therefore, the Platonic pronouncement on the title page is due to his successor as editor, not to Rheticus.

[2]Leopold Prowe, *Nicolaus Coppernicus* (Osnabrück, 1967; reprint of Berlin, 1883–84 ed.), I, 1, 291, omitted one "C" in this document's date, given in the Roman style. Despite its age and defects, Prowe's still remains the standard biography of Copernicus.

cine, he revolutionized astronomy, the study of which, in his opinion, invigorates the student's mind, gives him intellectual pleasure, directs his attention toward beautiful objects, and draws him away from vices. Besides conferring these mental, aesthetic, and moral benefits on its devotee, astronomy serves the community by regulating the calendar. According to Copernicus, astronomy is studied, not for the sake of astronomy, but to promote the welfare of the individual and society.[3]

When Copernicus said that astronomy or "mathematics is written for mathematicians," was he casting "a contemptuous side-glance at the masses?" Let us cast a direct glance at what he wrote:

Perhaps there will be babblers[4] who, although completely ignorant of mathematics, nevertheless take it upon themselves to pass judgment on mathematical questions and, badly distorting some passage of Scripture to their purpose, will dare find fault with my undertaking and censure it. I disregard them even to the extent of despising their criticism as unfounded. For it is not unknown that Lactantius, otherwise an illustrious writer but hardly a mathematician, speaks quite childishly about the earth's shape when he mocks those who declared that the earth has the form of a globe. Hence scholars need not be surprised if any such persons will likewise ridicule me. Mathematics is written for mathematicians.[5]

In the foregoing passage Copernicus did not express contempt for the masses, who in his time were not taught Hebrew, Greek, and Latin, and were therefore incapable of "badly distorting some passage of Scripture to their purpose." That ability was then confined to theologians. Copernicus correctly anticipated that there might be theologians so "completely ignorant of mathematics" or astronomy that they would presume to censure his geokinetic cosmology by misinterpreting the Bible. They would thereby become the contemporary counterparts of the Christian theologian Lactantius, who in antiquity poked fun at the upholders of the earth's sphericity. It was against such miseducated theologians thàt Copernicus directed his scorn, not at the uneducated common people.

The salutary warning prominently displayed on the title page of Copernicus' *Revolutions of the Heavenly Spheres* (1543) was addressed not only to non-mathematical theologians but also to all other potential readers untrained in geometry. Even the most cursory inspection of Copernicus' intricate masterpiece would persuade any serious student that he could not hope to cope with its closely reasoned demonstrations unless he had previously acquired the req-

---

[3] *Nikolaus Kopernikus Gesamtausgabe* (Munich & Berlin, 1944–49), II, 8–9.

[4] The Greek word *mataiologoi* used here by Copernicus was misequated with "exegetes" by Stillman Drake, *Discoveries and Opinions of Galileo* (Garden City, 1957), 180. In Paul's Epistle to Titus, 1:10, *mataiologoi* is translated as "vain talkers" (King James Bible). In his *Laws* (759c) Plato ordained that "from Delphi rules must be brought governing all matters of divinity, and Interpreters (exegetes) of these rules must be appointed." Plato's exegetes were no vain talkers: "everybody must consult the Interpreters and obey them" (*Laws*, 775a). Copernicus' *mataiologoi* were vain talkers or babblers, not exegetes.

[5] *Gesamtausgabe*, II, 6:27–35. In the same spirit Rheticus acknowledged in his *Preface to Arithmetic* that he was devoted to studies which "do not win the applause of the crowd" (Burmeister, I, 28).

uisite mathematical training. The Platonic injunction, "Learn geometry before attempting to read this book," was designed to attract qualified readers and discourage those unqualified. The latter group, far from being made the target of contempt, were given sound advice: don't waste your time or money.

Six years earlier the title page of Niccolò Tartaglia's *Nova Scientia* (Venice, 1537) presented an engraving which showed Euclid opening the door to a broad enclosure containing the mathematical sciences, including astronomy. Those students who have passed Euclid and these sciences may ascend to the loftier and more restricted enclosure of philosophy only if they are admitted by Plato, whose banner proclaims: "Nobody untrained in geometry may enter here." In this emphasis on geometry as a propaedeutic Copernicus' *Revolutions* was preceded by the *New Science* of Tartaglia. Has the latter ever been accused of despising the masses?

Did Copernicus ever make the statement about the masses which was recently attributed to him? For the answer to this question, let us turn to the biography of Copernicus published in 1654 by the French philosopher Pierre Gassend (1592–1655), who candidly avowed:

The little I knew about the man [Copernicus] would easily be learned by any reader of the books by him or others since at this time, a century after his death [in 1543], I had access to no other information about him nor to any other source.[6]

In an earlier biography of Copernicus[7] Gassend read that the astronomer

incurred the enmity of the Grand Master [of the Order] of Teutonic Knights because at the behest of the king [of Poland] he recovered the diocesan property unjustly seized by the Grand Master and restored it to the Church. Copernicus' other enemies were some courtiers and a certain Elbląg schoolmaster, who with melodramatic malevolence in the theater ridiculed Copernicus' opinion about the motion of the earth.[8]

Although Gassend's source, his slightly older contemporary Starowolski (1588–1656), correctly treated the Teutonic Knights and the Elbląg schoolmaster as entirely separate enemies of Copernicus, the Frenchman brought these foes of the astronomer together. Referring to the Knights as "unjust and powerful usurpers," Gassend imagined that

in a dignified manner Copernicus disregarded their threats and other tricks. In particular he paid no attention to the stratagem by which they incited[9] the schoolmaster of Elbląg to produce a public comedy ridiculing Copernicus, as Aristophanes once mocked Socrates, and with all [sorts of] jeers and gibes to

[6]Gassend, *Nicolai Copernici . . . vita* (Paris, 1654), 3–4.

[7]Prowe's reference to "Gassend, the first biographer of Copernicus" is a deplorable slip (I, 1, 46).

[8]Szymon Starowolski, *Scriptorum polonicorum . . . vitae* (Venice, 1627²), 158. The quoted material was not present in Starowolski's first ed. (Frankfurt am Main, 1625). Starowolski's two biographies of Copernicus were reprinted in *Zeitschrift für die Geschichte und Altertumskunde Ermlands*, IV (1867–69), 536–39 (cited hereafter as ZGAE).

[9]This misunderstanding of Starowolski was repeated by Prowe (I, 2, 233).

make the crowd hiss him on account of his theory concerning the motion of the earth.[10]

Gassend was so unfamiliar with the actual circumstances of Copernicus' life that he failed to recognize the name of the town in which the local schoolmaster's show made fun of Copernicus. In Latin and German the place is called "Elbing," and in Polish "Elbląg." Gassend, however, who knew Holland at first hand,[11] confused the scene of the anti-Copernican play with the Dutch village Elburg, even though Elbląg-Elbing is only a few miles away from Frombork-Frauenburg, the seat of Copernicus' Cathedral Chapter.[12]

Gassend's fertile inventiveness devised the further fantasy that "Copernicus' excellence, nevertheless, was so evident that it was rather the author of the comedy himself who was hissed[13] and meanwhile incurred the wrath of good men." Since Gassend himself acknowledged that he depended exclusively on printed sources, wherever he ventures on an excursion beyond them we must suspect that he is indulging his creative fancy, as in the present instance.[14]

As Copernicus' reaction to the Elbląg incident Gassend supposed that the astronomer "could have said what someone else once said."[15] Insufficiently attentive readers of Gassend failed to notice that the Frenchman did not attribute the remark in question to Copernicus. Of such inattentiveness the following is one example, which exerted an enormous and baneful influence on the subsequent biographical tradition of Copernicus: The Teutonic Knights

paid traveling actors and comedians, commissioning them to parody Copernicus and subject him to ridicule. It was easy to amuse the public by talking nonsense about a novel concept that was contrary to appearances and accepted ideas. The crowd came running to laugh and applaud. The buffoons took in lots of money, and repeated this show in town after town, even coming close to the astronomer's residence. Copernicus' friends, aroused to indignation, urged him to oppose these shows, all the more because the crowd gathered in great numbers and applauded this shameful parody. "Let them alone," Copernicus answered.[16]

---

[10]Gassend, *Copernici . . . vita*, 40.

[11]Joseph Bougerel, *Vie de Pierre Gassendi* (Paris, 1737), 37-64.

[12]Gassend's geographical error was not remarked by Prowe, who fully realized the Frenchman's shortcomings as a biographer of Copernicus: "The general situation of Prussia remained unknown to the learned mathematician living in far-off France. He is untrustworthy about many details connected with Copernicus' immediate surroundings" (I, 2, 327). Some specific blunders by Gassend were pointed out by Prowe (I, 1, 132, 370 71, 386; I, 2, 158, 278; II, 289).

[13]The typographical error "exhibitatus" was corrected to *exsibilatus* (100, misnumbered 110).

[14]Although Prowe himself corrected many errors committed by Gassend (see n. 12, above), he described his French predecessor as an "otherwise thoroughly reliable man" (I, 2, 234). According to Prowe (I, 1, 99), post-Gassend archival research "completely confirmed the reliability of Gassend's statements": "there is therefore no reason to doubt that the careful writer [Gassend] followed reliable sources also in the other parts of his biography" (I, 2, 502).

[15]Gassend, *Copernici . . . vita*, 40: ". . . idem posset. quod olim alius dicere."

[16]Jean Czynski, *Kopernik et ses travaux* (Paris, 1847), 67-68.

The astronomer continued with what, according to Gassend when read attentively, Copernicus "could have said," but in fact did not say: "I never wanted to please the people, since what I know is not approved by the people and what the people approve I do not know."

Despite Gassend's assertion that Copernicus could have said this, plainly coupled with the implication that he did not do so,[17] in the post-Gassend literature of the subject this utterance often became "Copernicus' famous statement." Yet it did not become famous enough to resist alteration. For instance, in the form which we encountered at the beginning of this article, the masses mock Copernicus, although the element of mockery was not present in Gassend. He referred to the gap between what the scholar knows and what the people approve. That gap has now turned into a difference between the desires of the two parties.

When Gassend asserted that Copernicus "could have said what someone else once said," did he know who that someone else was?

"I never yearned to please the masses since what pleased them was not understood by me, and what I knew was remote from their comprehension," said Epicurus (341–270 B.C.).[18] This was by no means a unique utterance on the Greek philosopher's part. Thus he emphasized his independence of thought and aloofness from the common man, for whose advantage he nevertheless strove, when he declared: "In the investigation of nature I would rather speak freely and prophesy what is beneficial to all mankind, even if nobody is likely to agree, than conform to [conventional] beliefs and thereby win constant praise coming from the masses."[19] As a personal reason for withdrawing from public life Epicurus asserted: "Neither relief from emotional disturbance nor worthwhile joy is obtained from . . . the respect and plaudits of the masses"; "the most absolute security comes from a tranquil life and avoidance of the masses."[20] A later Epicurean undoubtedly expressed his founder's attitude when he said: "The wise man . . . will conduct a school, but not so as to attract a crowd. And he will give readings in public, but not on his own initiative."[21]

Epicurus was the author of the saying attributed by Gassend to someone other than Copernicus. But Epicurus' saying was not quoted by the French philosopher in its original Greek form, which Gassend evidently had not encountered. Instead he repeated the somewhat altered version given by the Roman philosopher Seneca, who did not translate Epicurus literally.[22] The Greek thinker's statement "What I knew was remote from the comprehension of the masses" was modified by Seneca to read: "What I know is not approved by

---

[17]Here Gassend was read correctly by Prowe (I, 2, 234), who nevertheless made no effort to identify the author of our quotation.

[18]Epicuro, *Opere*, ed. Graziano Arrighetti (Turin, 1960), 437, n. 122; Cyril Bailey, *The Greek Atomists and Epicurus* (New York, 1964; reprint of Oxford, 1928 ed.), 225.

[19]Ed. Arrighetti, 145, n. 29; Cyril Bailey, *Epicurus* (Oxford, 1926), 111, n. XXIX.

[20]Ed. Arrighetti, 157, n. 81; 125, n. XIV; Bailey, *Epicurus*, 119, n. LXXXI; 98, n.XIV; *Greek Atomists*, 501.

[21]Ed. Arrighetti, 27–29; Bailey, *Epicurus*, 166; Diogenes Laertius, *Lives of the Philosophers*, X, 121b.

[22]*Letters*, 29, 10.

the people." Epicurus' emphasis on popular incomprehension was converted by Seneca into mass disapproval.[23]

Epicurus, ardent advocate of the simple life, avowed enemy of erotic pleasure, and pious worshipper of the gods, had long been ignorantly and vociferously reviled as a self-indulgent voluptuary, promiscuous profligate, and subversive atheist. Against these grotesque accusations Gassend had cautiously yet courageously defended the ancient Greek thinker in his *Epicurus' Life and Character* (Lyon, 1647).[24] In this work Gassend saw no reason to conceal the authorship of our quotation and ascribed it explicitly to Epicurus.[25] Yet seven years later in his biography of Copernicus Gassend evidently felt that it was prudent to conceal Epicurus' name and attribute our quotation to an anonymous "someone else."

Our quotation reveals the attitude toward the common people which was held by Epicurus, not by Copernicus. In order to understand the astronomer's actual attitude toward the common people, we shall now fulfill the promise made near the beginnning of this article and take a brief look at Copernicus' medical practice.

According to his biographer Starowolski,

In medicine Copernicus was honored as a second Aesculapius, even though with his entirely philosophical outlook he never craved a display [of adulation] on the part of the common people. For, as Tiedemann Giese, bishop of Chełmno [Kulm], writes about him elsewhere, he paid no attention to any nonphilosophical matters.[26]

Giese (1480–1550) was able to help overcome Copernicus' prolonged reluctance to publish the *Revolutions* because he was the astronomer's closest friend. Within two months after Copernicus' death on May 24, 1543, upon receiving printed copies of the *Revolutions* from Rheticus, on July 26 Giese wrote him a letter, urging him to try to get rid of the gross falsification which had been interpolated in the *Revolutions* while it was being edited in distant Nuremberg. Giese was aware, he told Rheticus, "of how highly Copernicus used to value your work and good will in aiding him."[27] Yet "in the *Revolutions'* preface" (where Copernicus described Giese as a "man who loves me very dearly"),[28] Giese pointed out to Rheticus, "your teacher [Copernicus] failed to mention you."[29]

[23]Gassend's quotation of Seneca's version, rather than Epicurus', indicates that the French philosopher was unacquainted with the original statement in Greek. This had been printed, without being attributed to Epicurus, in the *Melissa* of the 11th-century monk Antony (Zürich, 1546) as well as in two editions of John Stobaeus (Frankfurt, 1581; Geneva, 1609). Two mss, one in Naples, and the other in Florence, likewise contained the Greek original assigned to no author, whereas a Paris ms included it in the "Sayings of Epicurus." Gassend, however, knew as little about these three ms appearances of Epicurus' saying in Greek as he did about the three printed occurrences.

[24]Bernard Rochot, *Les travaux de Gassendi sur Épicure et sur l'atomisme 1619–58* (Paris, 1944), 126–33.

[25]Gassend, *De vita et moribus Epicuri* (The Hague, 1656²), 55.

[26]ZGAE, IV, 537.

[27]Burmeister, *Rhetikus*, III, 55.

[28]*Gesamtausgabe*, II, 3–4.

[29]Burmeister, III, 55.

Why did Copernicus omit Rheticus from the *Revolutions'* preface? This document was dedicated to Pope Paul III and honorifically named Nicholas Schönberg, cardinal of Capua, as well as Tiedemann Giese, bishop of Chełmno. It then referred to "not a few other very eminent and very learned men"[30] without recording their names. These anonymous friends of Copernicus included Rheticus. As a Protestant and professor of mathematics at Wittenberg University, the militant think-tank of the anti-papal Lutheran heresy, Rheticus would have been conspicuously out of place in the Roman Catholic team of pope, cardinal, and bishop. Rheticus knew this full well, and so did Giese. Yet it would have been most undiplomatic on Giese's part to say so bluntly in his letter to Rheticus, who had been his house guest for several weeks as Copernicus' companion.[31]

Consequently Giese had to find a palatable rationalization of Copernicus' omission of Rheticus' name from the *Revolutions'* preface. Copernicus did so, Giese told Rheticus in his letter of July 26, 1543, "not because he disregarded you, but on account of a certain apathy and carelessness (since he paid no attention to any nonphilosophical matters)."[32]

Copernicus' omission of Rheticus' name from the *Revolutions'* preface was recently characterized by Arthur Koestler as a "scandal" and "betrayal of Rheticus."[33] The latter's reference to Copernicus in the dedication of one of his books was said by Koestler to be the "last affirmation of the pupil's loyalty."[34] Rheticus' dedication is dated on the Ides of August,[35] 1542.[36] Fifteen years later, in another dedication, Rheticus publicly referred to "Nicholas Copernicus, the Hipparchus of our age, who is never praised enough."[37]Besides elevating Copernicus to the rank of Hipparchus, often regarded as the most original ancient astronomer, in this same dedication Rheticus described his attitude toward Copernicus by saying: "I have always cherished, esteemed, and honored him not only as a teacher but also as a father."[38] How well Koestler knew Rheticus' career may be judged from Koestler's appointment of Rheticus to be a professor, not at the University of Wittenberg, but "at the University of Nuremberg,"[39] which never existed. Rheticus also studied at the University of Nuremberg (according to Koestler, 154, 156), as well as at the University of Göttingen, where he was also a professor—at an institution that was inaugurated in 1737,[40] more than a century and a half after Rheticus' death in 1574!

In Giese's letter to Rheticus of July 26, 1543 is found the source of the biographer Starowolski's statement, quoted above, that Copernicus "paid no attention to any nonphilosophical matters." "Elsewhere," let us recall, Starowolski encountered Giese's report that "In medicine Copernicus was

---

[30]*Gesamtausgabe*, II, 4: 5–6.

[31]Edward Rosen, *Three Copernican Treatises* (New York, 1959²), 109.

[32]Burmeister, III, 55. Giese's parenthesis (*"ut erat ad omnia quae philosophica non essent minus attentus"*) was omitted from the translation into German in Burmeister (III, 58), who merely reprinted Franz Beckmann's version in ZGAE, 1861–63 (II), 352.

[33]Koestler, *The Sleepwalkers* (New York, 1968; reprint 1959), 172.

[34]Koestler, 174.   [35]August 13, not 15, as in Koestler, 174.   [36]Burmeister, III, 51.

[37]Burmeister, III, 138.   [38]Burmeister, III, 139.   [39]Koestler, 174.

[40]Götz von Selle, *Universität Göttingen, Wesen und Geschichte* (Göttingen, 1953), 20.

honored as a second Aesculapius, even though with his entirely philosophical outlook he [Copernicus] never craved a display [of adulation] on the part of the common people." Unfortunately the document in which Giese recorded Copernicus' popular reputation as a doctor has not been recovered, by contrast with Giese's letter to Rheticus of July 26, 1543 about Copernicus' alleged inattention to nonphilosophical matters.

In his characteristic way Gassend enlarged upon the Giese-Starowolski remark about Copernicus being honored as a second Aesculapius, the patron of physicians who became the god of healing in Greco-Roman antiquity.[41] The Giese-Starowolski reference to the second Aesculapius "is properly understood to mean," said Gassend, that Copernicus

had a thorough knowledge of certain special medicaments, prepared them himself, and administered them successfully in dispensing them to the poor, who therefore revered him as a sort of divinity. For otherwise the general practice of medicine was contrary to his way of life.[42]

In an earlier passage of his biography Gassend had explained that Copernicus resolved to devote himself to three principal tasks, of which one was "with whatever skill he had in medicine, never to fail the poor who besought his help."[43] In treating his colleagues Copernicus fulfilled the aforementioned promise he had made to his Chapter in 1501. In ministering to the poor, Copernicus heeded the Hippocratic *Precepts* (VI): "Take account of [your patient's] luxury and property. Sometimes donate [your services]."

The statement that "Copernicus was honored as a second Aesculapius" by the common people is historically trustworthy since it emanated from Giese, who was an intimate friend of Copernicus. Even though the astronomer, according to Giese, "never craved a display [of adulation] on the part of the common people," he was evidently on good terms with them. He cast no contemptuous side-glances at them. As a physician, he desired what they desired: sound public health.

True, his geokinetic cosmology was derided in a theater. But that happened only once in Elbląg. Copernicus' arguments in favor of his revolutionary astronomy were inaccessible to the masses, who had not received the necessary preliminary instruction in mathematics and in any case knew no Latin, the language in which scientific treatises were then written.

Such a linguistic barrier between a scholar and his public did not exist for Epicurus, who wrote in Greek for Greek-speaking readers. In fact, one of his followers was certain that Epicurus' gods, who enjoyed talking to one another, spoke Greek.[44] "I never yearned to please the masses since what pleased them was not understood by me, and what I knew was remote from their comprehension" was an expression of Epicurus', not Copernicus' attitude toward the common people.

[41]Emma J. and Ludwig Edelstein, *Asclepius* (Baltimore, 1945), II, 95.

[42]Gassend, *Copernici . . . vita*, 39 not 59, as in Prowe, I, 1, 294.   [43]Gassend, 7.

[44]Philodemus, *On the Gods*, Book III, column 14, lines 6–8: "We must believe that the gods' language is Greek or not very different therefrom" (*Abhandlungen der k. Preussischen Akademie der Wissenschaften* [phil.-hist. Klasse, 1916], n. 4, p. 37; n. 6, p. 50).

9

Copernicus' Earliest Astronomical Treatise
(with Erna Hilfstein)

When Copernicus in 1508 prepared for the printer his Latin translation of *Letters* by the 7th century Byzantine writer Theophylactus Simocatta,[1] his manuscript began with a brief dedication "To the Most Reverend Lucas, bishop of Varmia," by Nicholas Copernicus.[2] He may not have foreseen that his dedication, printed in smaller type, would be placed after an introductory poem, printed in much larger type. But, whatever fonts were chosen by the publisher, Copernicus named himself as the translator, and so he appeared in print.

The same cannot be said about his earliest astronomical treatise, from which he withheld his name.[3] When this treatise was listed in an inventory completed on 1 May 1514, the entry referred to the "theory of an [unnamed] author who asserts that the earth moves while the sun stands still."[4] Why did Copernicus reveal his name as the translator of Theophylactus, but conceal his name as the author of the geokinetic theory? From his translation dedicated to the bishop Copernicus had little to fear, apart from the love letters, which he toned down. But his treatise proclaiming the motion of the earth made it a solar planet, that is, a heavenly body. Copernicus' alteration of the earth's cosmic status, from a stationary non-heavenly body to a revolving heavenly body, dissolved the traditional distinction between heaven and earth. Any human being born on the earth is by that very fact already in the Copernican heaven. Be a good boy or girl, and wait until death before entering heaven? Everybody is already there at birth. This inescapable theological consequence of the earth's annual

---

[1] Nicholas Copernicus, *Theophilacti scolastici Simocati epistole morales, rurales et amatorie interpretatione latina* (Cracow: Johannes Haller, 1509). English translation in *Nicholas Copernicus Complete Works*, III, ed. Paweł Czartoryski, translation and commentary by Edward Rosen with the assistance of Erna Hilfstein (Warsaw/Cracow: Polish Scientific Publishers, 1984), pp. 3—71; hereafter NCCW III.

[2] *Ibid.*, fol. a$_{iiii}$recto: *Ad reverendissimum dominum Lucam episcopum warmiensem Nicolai Copernici epistola.*

[3] NCCW III, pp. 75—76.

[4] Leszek Hajdukiewicz, *Biblioteka Macieja z Miechowa* (Wrocław: Ossolineum, 1960). p. 218, no. 189: *sexternus theorice asserentis terram moveri, solem vero quiescere.*

revolution around the sun was fraught, in that age of ominous religious unrest, with grave danger for the propounder. No wonder he prudently withheld his name!

He withheld his name, but not his thoughts. It is not true "that he concealed his theories" throughout "the entire period that had elapsed since his first discovery of the heliocentric theory."[5] On the other hand, he did not rush pell-mell into print. Instead he steered a middle course. Avoiding complete silence on one side, and unrestricted publication on the other side, he distributed handwritten (anonymous) copies of his treatise to a few trusted professional friends.

As is shown by the entry in the 1514 inventory, Copernicus' earliest astronomical treatise did not reveal its author's name. Yet sixty-one years later the greatest living astronomer Tycho Brahe (1546—1601) learned that the author was Copernicus. The original manuscript written by Copernicus with his own hand (and without his name being disclosed) was given to Brahe by the personal physician of the German prince who was crowned King of the Romans (and future Holy Roman Emperor) in Regensburg in 1575. On 1 November of that year at the coronation ceremonies Dr. Thaddeus Hájek (1525—1600) met Brahe, and discovered that the latter had a very high regard for Copernicus. His only disciple, Georg Joachim Rheticus, who had died on 4 December of the previous year,[6] had bequeathed a substantial part of his library to his fellow-physician Hájek. What greater service to astronomy could Hájek perform than to pass on to Brahe, the greatest living astronomer, the earliest astronomical treatise by the greatest astronomer of the previous generation? In 1539 Rheticus had arrived in Frombork with an armful of valuable books for his new master, who in turn gave Rheticus gifts in exchange, including the anonymous treatise. But of course Rheticus knew who the author was, and so informed his legatee Hájek, who so informed his friend Brahe. In his *Astronomiae instauratae progymnasmata*, Part 2, the printing of which was begun in 1588 and not finished until 1602, the year after the death of Brahe, he related that

A certain little treatise (*Tractatulus*) by Copernicus concerning the hypotheses which he formulated, was presented to me in handwritten form some time ago at Regensburg by that most distinguished man Thaddeus Hájek, who has long been my very close friend. Subsequently I sent the treatise to certain other astronomers in Germany.

---

[5] Noel M. Swerdlow, "The Holograph of *De Revolutionibus* and the Chronology of Its Composition," *Journal for the History of Astronomy*, 1974, 5: 191/3—5. Two years after disseminating the misinformation that Copernicus "concealed his theories" throughout "the entire period that had elapsed since his first discovery of the heliocentric theory," Swerdlow wrote about the "readers to whom Copernicus sent this brief description of his new planetary theory;" Swerdlow, "Pseudodoxia Copernicana," *Archives internationales d'histoire des sciences*, 1976, 26: 116/11↑—10↑.

[6] Karl Heinz Burmeister, *Georg Joachim Rheticus, 1514—1574*, I (Wiesbaden: Guido Pressler Verlag. 1967). p. 177/11—12.

I mention this fact to enable the persons into whose hands the manuscript comes to know its provenience.[7]

When Brahe's assistants were preparing the copies to be sent out, he turned over to them Copernicus' previously untitled treatise, for which he devised the following title:

> *Nicolai Copernici de hypothesibus motuum caelestium*
> *a se constitutis commentariolus*
> (Nicholas Copernicus' little commentary on the hypotheses
> formulated by himself for the heavenly motions)

The last word in the title he devised was for the moment forgotten 'by Brahe when years later he wrote his *Progymnasmata*, Part 2, which instead used the synonym *Tractatulus* (little treatise). This designation failed to gain the widespread recognition accorded to *Commentariolus* (little commentary), which became the standard short title of the treatise left without title by Copernicus.[8]

The year before the *Commentariolus* was acquired by Brahe, he addressed the entire faculty and student body of the University of Copenhagen early in September 1574. In those lectures he spoke of Ptolemy's *hypotheses ab ipso constitutas* (hypotheses formulated by himself) and Copernicus' *hypothesibus aliter constitutis* (hypotheses formulated differently).[9] Copernicus himself however, in the *Commentariolus* cautioned his readers not to suppose that his hypotheses were formulated gratuitously, following those propounded by the Pythagoreans.[10] In the *Revolutions* he named individual Pythagoreans (Ecphantus, Hicetas, Philolaus), and expounded their policy of public relations.[11] The Roman Catholic Sacred Congregation of the Index on 5 March 1616 condemned "the Pythagorean doctrine of the earth's mobility and the sun's immobility, a doctrine which is false and completely contrary to divine Scripture and which Nicholas Copernicus teaches in his" *Revolutions*.[12] Thus it is clear that Noel M. Swerdlow errs when he states that "Copernicus... accuses them [the Pythagoreans], justly I think, of not knowing what they were talking about."[13]

"That Copernicus left the treatise [*Commentariolus*] untitled... is by no means certain... He may in fact have called it something like *De hypothesibus*

---

[7] *Tychonis Brahe dani opera omnia*, II (Copenhagen, 1913—1929), p. 428/34—40; hereafter TB.

[8] Hajdukiewicz, *Biblioteka*, p. 384/8—10, believes that the treatise had the title *De hypothesibus motuum coelestium a se constitutis Commentariolus* since its inception.

[9] TB, I, p. 149/24—25, 28.

[10] NCCW, III, p. 82/15—16.

[11] NCCW, II, pp. 3/15—19; 4/46; 5/1, 3: 25/26—32.

[12] *Ibid.*, p. 342/11—13.

[13] Noel M. Swerdlow, "The Derivation and First Draft of Copernicus's Planetary Theory: A Translation of the Commentariolus with Commentary," *Proceedings of the American Philosophical Society*, 1973, p. 439/right/11—15.

motuum caelestium," according to that same misinformant,[14] who is obviously unaware that the term *hypothesibus* is never used in the *Commentariolus.* When this expression later became part of Copernicus' vocabulary, in the *Revolutions* it occurred mainly as a synonym for "principle" and "assumption": "... astronomy's principles (*principia*) and assumptions (*assumptiones*), called 'hypotheses' by the Greeks."[15]

In the *Commentariolus* Copernicus introduced seven[16] "postulates [*petitiones*], which are called axioms [*axiomata*]."[17] These are "incorrectly called *axioms* since they are hardly self-evident," says Swerdlow,[18] thus accusing Copernicus, a student from 1491 to 1503 in three outstanding universities whose language of instruction was Latin, of misunderstanding the word *axiom.* This term was avoided by Euclid, whose geometry provides the earliest extant example of the axiomatic method. But in his *Commentary on the First Book of Euclid's Elements* the famous philosopher Proclus (c. 410—485) discussed "what are generally called indemonstrable axioms, inasmuch as they are deemed by everybody to be true and no one disputes them."[19] *Commentariolus'* axioms are different, since its readers are asked to grant (*concendantur*) the "postulates which are called axioms." On the other hand, *Commentariolus'* axioms correspond to the axioms defined in Aristotle's *Posterior Analytics* as "the primary propositions from which a proof proceeds."[20]

To designate these primary propositions, *Commentariolus* used the term *petitiones,* which had also been employed by a brief medieval treatise *On Floating Bodies.* Its six primary propositions (*petitiones*) were intended as true statements about the real world, rising in an unmistakable crescendo:

1. No body is heavy in relation to itself.
2. Every body weighs more in air than in water.

6. If equal [volumes of] bodies are equal in weight, their specific gravities are said to be equal.[21]

By the same token, *Commentariolus'* seven *petitiones* mount in a magnificent ascent. Postulate 1 ("There is no one center")[22] undermined the principle

---

[14] *Ibid.,* p. 423/right/17—22.

[15] NCCW, II, p. 7/40—41.

[16] Leopold Prowe, *Nicolaus Coppernicus,* I² (Osnabruck: Otto Zeller, 1967; reprint of the 1883—1884 ed.), p. 291/14—15, omitted the heading over the seventh postulate; he corrected this in II, 197, n. */last 5 lines; hereafter Prowe.

[17] NCCW, III, 81/27—28.

[18] Swerdlow, "Derivation," p. 437/left/7↑—6↑.

[19] Proclus, *A Commentary on the First Book of Euclid's Elements,* translated by Glenn R. Morrow (Princeton, 1970), p. 152/3—5.

[20] Aristotle, *Posterior Analytics,* I, 10; 76b/4—15.

[21] *The Medieval Science of Weights,* eds. E. A. Moody and M. Clagett, 2nd printing (Madison, 1960), p. 42.

[22] Postulate 1 is turned into its opposite by Prowe, I², 290/8--9: "there is only one center" (*gibt es nur einen Mittelpunkt*; stress on "einen" is Prowe's).

of concentricity, which had required all the celestial spheres to share a common center, the earth. Postulate 2 proclaimed that "the earth's center is not the universe's center," which "is near the sun" (Postulate 3). The distance from the earth to the sun is imperceptible in comparison with the remoteness of the stars (Postulate 4). The earth's actual daily rotation accounts for the stars' apparent daily rotation (Postulate 5). The apparent annual cycle of the sun's movements is caused by the earth's real revolution around the sun (Postulate 6), which also makes the planets seem to retrogress intermittently (Postulate 7). These seven postulates or axioms were not set forth by Copernicus as self-evident, the requirement anachronistically foisted on them by our misinformant. These axioms, far from being self-evident, would startle and shock his contemporaries, as Copernicus was painfully aware. His fear of what they might do to vent their disapproval of him induced him to withhold his name as author of the *Commentariolus*.

This treatise is charged by our misinformant with "the logical error of stating Postulates 2, 4, 5, and 7 as postulates rather than deductions from Postulates 3 and 6." [23] He is more suitably to be charged with the anti-historical error of assimilating Copernicus to Giuseppe Peano (1858—1932), who set out "to enunciate as postulates only those that cannot be deduced from others which are simpler" and "to reduce them to the smallest number." [24]

Ptolemy, the greatest of the ancient Greek astronomers, had taught that

all the spheres are closer to the earth than that of the fixed stars... [It] is carried around by the primary [daily] motion from east to west about the poles of the equator, but also has a proper motion in the opposite direction about the poles of the sun's ecliptic circle. [25]

These two motions attributed to the starry sphere by Ptolemy are taken away from it and assigned to the earth by *Commentariolus*. Hence its Postulate 5 announces: *firmamento immobili permanente ac ultimo caelo* (the firmament and highest heaven abide unchanged). [26] These six simple Latin words are twisted by Swerdlow into "the sphere of the fixed stars remains immovable and the highest heaven." [27] He then strives to support his ungrammatical blunder by arguing that Postulate 5's "sphere of the fixed stars is the outermost heaven because a sphere of diurnal rotation is no longer necessary." [28] But a sphere of diurnal rotation is always necessary, since diurnal rotation is an astronomical fact. For the *Commentariolus* that necessary sphere of diurnal rotation was the earth. For Ptolemy, the diurnal rotation has been performed by the starry sphere, which moved, whereas *Commentariolus'* did not.

---

[23] Swerdlow, "Derivation," p. 46/19—20.

[24] Giuseppe Peano, *Opere scelte* (Rome, 1957—1959), III, 116/9↑—8↑, 119/5↑—4↑.

[25] *Ptolemy's Almagest*, translated and annotated by Gerald J. Toomer (London: Duckworth & Co., 1984), VIII, 3, p. 404) 12↑—9↑; IX, 1, p. 419/11—12; hereafter Ptolemy.

[26] NCCW, III, 81/4↑—3↑.

[27] Swerdlow, "Derivation," p. 436/Fifth Postulate/3—4.

[28] *Ibid.*, p. 438/right/#5/5—7.

Commentariolus analyzed a planet's

second anomaly. in which the planet is seen sometimes to retrograde and often to become stationary. This second anomaly happens by reason of the motion, not of the planet. but of the earth.[29]

Postulate 7, for the first time in the history of mankind. rejected the pre-Copernican view that the second anomaly was a planet's actual motion. The predecessors of Copernicus believed that a planet moved eastward at a constantly varying rate. Gradually it slowed down until it hardly moved at all, at its stationary point. Reversing its direction. it proceeded in retrogradation westward until its second station. There it resumed its regular eastward journey and repeated this looping pattern over and over.

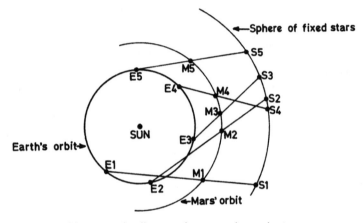

(Mars seen in direct and retrograde motion)

Postulate 7, however, explained these planetary loops — the second anomaly — as an optical effect, not a physical phenomenon. Consider the planet Mars as an example. In approximately two years it revolves around the sun, the circuit traversed by the earth on an inner track in one year. The observer on the earth at E_1 sees Mars in M_1 at S_1 in the sphere of the stars; from E_2, M_2 is seen at S_2; from E_3, M_3 is seen at S_3, the stationary point; from E_4, M_4 is seen to have retrogressed to S_4; from E_5, M_5 is seen to have resumed its direct journey to S_5. This quantum leap in comprehension of the universe was accomplished by Postulate 7. It provided a simple, clear, and decisive statement that the earth, a planetary satellite revolving around the sun, stationary in the center of the universe, is responsible for the optical effect of loops in the outer planets. No such idea ever emerged, or could have emerged. from pre-Copernican geocentrism, convinced that the earth was stationary in the center of the universe. Swerdlow displays his incomprehension of this spectacular advance by claiming that "Copernicus has raised no objection to Ptolemy's representation of the

---

[29] NCCW. III. p. 86/18—20.

second anomaly"[30] or planetary loops. Why object to what has already been abolished? Why whip a dead horse? In addition, Swerdlow refers to Copernicus' assertion of the heliocentric theory "giving (correctly) as the only evidence for his assertion the equivalence of heliocentric to geocentric planetary theory."[31] Copernicus' heliocentrism explained the observed planetary loops as an optical effect produced by planet earth revolving around the central sun. On the other hand, the pre-Copernican geocentric planetary theory regarded the planetary loops as physical phenomena executed by the planets and viewed from a central, stationary earth. "Equivalence of heliocentric to geocentric planetary theory"? This alleged equivalence given by Copernicus "as the only evidence" for his heliocentrism? When and where did he ever do so? Here our ordinarily loquacious misinformant is strangely silent. After conceiving the heliocentric astronomy, Copernicus never for one moment swerved from his unalterable opposition to geocentrism as irreconcilable with his new heliocentric system. Thinking mankind long ago made up its mind that Copernicus' explanation of the planetary loops dealt the death blow to the geocentric planetary theory. His treatment of the loops as an optical effect has been repeated in countless introductions to and histories of astronomy all over the world for many centuries and in numerous languages. His account was accepted as decisive proof of the correctness of geokineticism. Yet according to Swerdlow, the only evidence given by Copernicus for his assertion of heliocentrism is its equivalence to geocentric planetary theory, which it overthrew after a reign of fourteen hundred years.

That overthrow was not accomplished by Copernicus alone. A powerful ally was that great Copernican, Galileo Galilei (1564—1642). In his magnificent *Dialogue Concerning the Two Chief world Systems — Ptolemaic and Copernican* (Florence, 1632), for which he was condemned to life imprisonment by the Roman Catholic church, Galileo demonstrated the mutual incompatibility, not the equivalence, of the heliocentric and geocentric planetary theories.[32] With his characteristic dramatic skill Galileo has a spokesman ask:

With all the proofs in the world what would you expect to accomplish in the minds of people who are too stupid to recognize their own limitations?[33]

In their newest book dealing with Copernicus' mathematical astronomy, Swerdlow-Neugebauer remark:

Little service has been done to his [Copernicus'] reputation by the common biographical tradition that he had thoroughly proved his case and merely feared that the rest of the world would be too stupid to understand.[34]

---

[30] Swerdlow, "Derivation," p. 437/right/30—31.

[31] *Ibid.*, p. 425/left/10—12.

[32] Galileo Galilei, *Dialogue Concerning the Two Chief World Systems — Ptolemaic & Copernican*, translated by Stillman Drake (Berkeley and Los Angeles, 1962), pp. 322—327.

[33] *Ibid.*, p. 327/5↑—3↑.

[34] Noel M. Swerdlow and Otto Neugebauer, *Mathematical Astronomy in Copernicus's De Revolutionibus* (New York/Berlin: Springer-Verlag), pp. 20/last-21/3; hereafter Swerdlow-Neugebauer.

No such judgment about mankind was ever expressed by Copernicus. "The common biographical tradition" that Copernicus felt such a fear about human stupidity is adduced by Swerdlow-Neugebauer, who attach 56 footnotes to their biography of Copernicus without offering a single example of this alleged "common biographical tradition." Does it exist? Did it ever exist?

A planet does not actually move back and forth. It is seen doing so only because the observer is situated on the earth making wide swings around the sun. In like manner, as concerns the moon, Ptolemy pointed out that

the straight line drawn from the center of the earth... through the center of the moon to a point on the ecliptic... determines the true position [of the moon]... The line drawn from some point on the earth's surface, that is, the observers point of view, to the moon's center... determines its apparent position. Only when the moon is in the observer's zenith do the lines from the earth's center and the observer's eye through the moon's center to the ecliptic coincide. But when the moon is displaced from the zenith position in any way whatever, the directions of the above lines become different, and hence the apparent position cannot be the same as the true.[35]

This difference between the moon's true and observed position is known as its parallax. A Latin equivalent of this Greek term was *commutatio* (commutation).[36] Copernicus decided to apply the term "motion in commutation" to his explanation of the planetary loops:

I say that the motion in parallax [commutation] is nothing but the difference by which the earth's uniform motion exceeds their motion, as in the cases of Saturn, Jupiter, and Mars, or is exceeded by it, as in the cases of Venus and Mercury.[37]

As the earth swings around the sun in the course of a year, a star observed on two nights six months apart might be expected to show a parallactic displacement due to the vast distance between those positions of the earth on opposite sides of the sun. But no such annual stellar parallax is visible to the naked eye. In the pre-Copernican astronomy, whose earth was central and stationary, this question of stellar parallax never arose. Thus, Ptolemy's *Planetary Hypotheses*, estimating the distances outward from the central earth, declared with regard to the outermost planet that "the greatest distance of Saturn, which is adjacent to the sphere of the fixed stars, is 19,865 earth-radii."[38] But when Copernicus shifted the earth out of the universe's center onto the third planetary track, how could the starry sphere continue to remain adjacent to Saturn's without displaying annual stellar parallax? He provided the answer in *Commentariolus'* Postulate 4:

---

[35] Ptolemy, IV, 1, p. 173/15↑—6↑.

[36] *Nicolai Copernici opera omnia*, II (Warsaw/Cracow: Polish Scientific Publishers, 1975), Bk. IV, 2; p. 176/24.

[37] NCCW, II, 228/16—18.

[38] Bernard R. Goldstein, "The Arabic Version of Ptolemy's *Planetary Hypotheses*," *Transactions of the American Philosophical Society* (1967), 57, p. 7/right/19↑—17↑.

The ratio of the earth's distance from the sun to the height of the firmament is so much smaller than the ratio of the earth's radius to its distance from the sun that the distance between the earth and the sun is imperceptible in comparison with the loftiness of the firmament.[39]

No annual stellar parallax is visible to the naked eye in Copernicus' geokinetic astronomy because the starry sphere is no longer adjacent to Saturn's, but is now enormously remote. Gone forever was Ptolemy's compact cosmic jewel box, with its shining stars some 20,000 earth-radii away from us. The Copernican "heavens are immense... and present the aspect of an infinite magnitude"; "the universe is... similar to the infinite."[40] In these immense Copernican heavens the search for annual stellar parallax went on nearly three centuries until the requisite improved observational technology found it. Uninformed people then said that at last Copernicus was proved right. But that proof did not have to wait for additional confirmation through annual stellar parallax, decisive proof having been provided from the start by planetary parallax. Our misinformant, it will be recalled, remarked about Copernicus' assertion of the heliocentric theory that he gave

(correctly) as the only evidence for his assertion the equivalence of heliocentric to geocentric planetary theory and the additional sense of the heliocentric representation of the second anomaly and the order of the planetary spheres.[41]

Postulate 4 replaced Ptolemy's cosmic jewel box by Copernicus' immense heavens. How can two such contrasting conceptions of cosmic space be regarded as equivalent? Our misinformant's weird verbiage ("the additional sense of the heliocentric representation of the second anomaly and the order of the planetary spheres") may be intended to mask Copernicus' spectacular success with regard to the planetary loops and the arrangement of the planets in their proper order. Ptolemy had said that

none of the stars [and planets] has a noticeable parallax (which is the only phenomenon from which the distances can be derived).[42]

On the other hand,

in Copernicus' theory the radius of the earth's orbit is the common measure of the radii of all the planetary orbits. It is this assumption... that allows the distances of the planet from the mean sun to be determined in what is often called the Copernican System,

according to Swerdlow-Neugebauer.[43] If it is not called the Copernican System, what then is it called?

---

[39] NCCW, III, 81, #4.

[40] NCCW, II, 13/32—33; 26/38—39.

[41] Swerdlow, "Derivation," p. 425/left/10—15.

[42] Ptolemy, IX, 1; p. 419/4↑—3↑.

[43] Swerdlow-Neugebauer, p. 322, #8/5—9.

10

Copernicus on the Phases and the Light of the Planets

What was new and what was old in Copernicus' opinions about the phases and the light of the planets? Did he regard the planets as opaque bodies, which in certain positions should show phases like the moon's? Or did he view the planets as self-luminous, like the sun? Or did he think of the planets as transparent throughout, so that the sunlight passing through them illuminated the whole visible disk, leaving no areas dark and thereby eliminating the possibility of phases?

A convenient starting-point in our effort to answer the foregoing questions will be the single most spectacular achievement in the long history of computational astronomy, namely, the discovery of the planet Neptune through the perturbations which it produced in the motion of Uranus [1]. At a meeting of England's Royal Astronomical Society on November 13, 1846, the Astronomer Royal, Sir George Biddell Airy (1801—1892), took the floor to defend himself against severe censure of his conduct in the exciting events which had culminated in the discovery of Neptune [2].

In the course of his *Account of Some Circumstances Historically Connected with the Discovery of the Planet Exterior to Uranus*, Airy praised the French theoretical astronomer Urbain-Jean-Joseph Le Verrier (1811—1877) for "the firmness with which he proclaimed to observing astronomers, «Look in the place which I have indicated, and you will see the planet well». Since Copernicus declared that, when means should be discovered for improving the vision, it would be found that

---

[1] Morton G r o s s e r, *The Discovery of Neptune*. Cambridge, Mass. 1962, pp. 49—57, 69, 75—76, 78—90, 92—123.

[2] Airy was recently defended in a paper of a later Astronomer Royal, Sir Harold Spencer Jones; see: Sir Harold Spencer J o n e s, *G. B. Airy and the Discovery of Neptune*. "Nature", CLVIII, 1946, pp. 829—830; reprinted in: "Popular Astronomy", LV, 1947, pp. 312—315.

Venus had phases like the Moon, nothing (in my opinion) so bold, and so justifiably bold, has been uttered in astronomical prediction".

Airy's oral reference to a prediction by Copernicus (1473—1543) that the phases of Venus would be detected by improved vision was promptly printed in the "Memoirs of the Royal Astronomical Society". But in the printed version Airy added the following footnote to his statement about Copernicus' prediction: "I borrow this history from Smith's *Optics*, sect. 1050. Since reading this «Memoir», I have, however, been informed by Professor De Morgan that the printed works of Copernicus do not at all support this history, and that Copernicus appears to have believed that the planets are self-luminous" [3].

Airy's footnote soon came to the attention of Alexander von Humboldt (1769—1859), the illustrious scientist whose name is now borne by the university which was founded principally by his brother [4]. At the time when Airy adverted to Copernicus' supposed prediction, Humboldt was engaged in writing his last and most famous work, into which he introduced the following passage:

"Whether Copernicus p r e d i c t e d the necessity of a future discovery of the phases of Venus, as is asserted in Smith's *Optics*, sect. 1050, and repeatedly in many other works, has recently become altogether doubtful, from Professor De Morgan's strict examination of the work *De Revolutionibus*, as it has come down to us" [5].

By inserting this passage in his *Kosmos*, which was repeatedly published in the original German as well as in numerous translations, Humboldt rendered the history of science a valuable service. That service would have been even more valuable had Humboldt indicated that the volume containing Airy's footnote also included De Morgan's "strict examination".

Augustus De Morgan (1806—1871), the immensely learned professor of mathematics at University College, London, had once been Airy's pupil at Cambridge, and later became his intimate friend. When the Royal Astronomical Society met on June 11, 1847, some seven months after Airy had told it about Copernicus' supposed prediction, it heard from De Morgan, who had recently been elected a secretary of the Society, a report *On the Opinion of Copernicus with Respect to the Light of the Planets*:

---

[3] "Memoirs of the Royal Astronomical Society", XVI, 1847, p. 411; also: "Monthly Notices of the Royal Astronomical Society", VII, 1847, p. 142.

[4] *Die Humboldt-Universität, Gestern — Heute — Morgen.* Berlin 1960, p. 18.

[5] Alexander von H u m b o l d t, *Kosmos.* Berlin 1850, vol. 3, p. 538. Citing "the letter from Adams to the Rev. R. Main on September 7, 1846". But John Couch Adams' letter to Robert Main, Airy's chief assistant at the Greenwich Observatory, did not refer to Copernicus. That reference was made by Airy, not by Adams.

"The common story is, that Copernicus, on being opposed by the argument that Mercury and Venus did not shew phases, answered that the phases would be discovered some day. The first place in which I find this story is in Keill's *Lectures*. It is also given by Dr. Smith, in his well-known treatise on *Optics*, by Bailly, and by others. But I cannot find it mentioned either by Melchior Adam or Gassendi, in their biographies of Copernicus; nor by Rheticus, in his celebrated *Narratio* [6], descriptive of the system of Copernicus; nor by Kepler, nor by Riccioli, in their collections of arguments for and against the heliocentric theory; nor by Galileo, when announcing and commenting on the discovery of the phases; and, what is most to the purpose, Müller, in his excellent edition of the great work of Copernicus, when referring to the discovery of the phases of Venus, as made since, and unknown to, Copernicus, does not say a word on any prediction or opinion of the latter. This story may then be rejected, as the gossip of a time posterior to Copernicus" [7].

This "gossip" was presumably started by John Keill (1671—1721), the excessively zealous proponent of Newton's priority over Leibniz in the invention of the calculus. In 1718, when Keill published the lectures which he gave as Savilian professor of astronomy at Oxford University, he emphasized the gossip by displaying *The Prophecy of Copernicus* as a marginal note alongside the following remarks:

"It was objected to him [Copernicus] that if the motions of the planets were such as he supposed them to be, then Venus ought to undergo the same changes and phases as the moon does. Copernicus answered that perhaps the astronomers in after-ages would find that Venus does really undergo all these changes. This prophecy of Copernicus was first fulfilled by that great Italian philosopher Galileo who, directing his telescope to Venus, observed her appearances to emulate the moon, as Copernicus had foretold" [8].

Keill's unsubstantiated gossip was repeated twenty years later by Robert [9] Smith (1689—1768), Plumian professor of astronomy at Cambridge University: "When Copernicus revived the ancient Pythagoric system, asserting that the earth and planets moved round the sun in the center of their orbits, the Ptolemaics objected, if this were true, that the phases of Venus should resemble those of the moon. Coperni-

---

[6] Translated into English by Edward Rosen, see: *Three Copernican Treatises,* 2nd ed. New York 1959, London 1959, pp. 107—196.

[7] "Monthly Notices of the Royal Astronomical Society", VII, 1847, pp. 290—291.

[8] John Keill, *Introductio ad veram astronomiam.* Oxford 1718, p. 194; English translation: *An Introduction to the True Astronomy.* London 1721, p. 163.

[9] Not "Thomas", as in: A. v. Humboldt, *op. cit.,* vol. 2, p. 362; this slip was corrected in: idem, *op. cit.,* vol. 5, p. 1289.

cus replied, that some time or other that resemblance would be found out" [10].

In the next generation Keill's gossip crossed the Atlantic Ocean with the eminent botanist José Celestino Mutis (1732—1808), a Spaniard by birth who founded at Bogotá in 1803 the first astronomical observatory in the Western Hemisphere. His voice was the earliest publicly to espouse Copernicanism in the New World, to the dismay of the Roman Catholic clergy. In the course of his eloquent reply to their attack on him, Mutis declared in 1774:

"The astronomers contemporary with Copernicus argued against him by saying that if his system were true, Venus should be observed crescent-shaped and less than full. Copernicus admitted that this should happen, and that the absence of this observation was due to the astronomers not having found the means of perfecting vision, a prophecy which came to be fulfilled in Galileo's time through the most fortunate invention of the telescope" [11]. Although Mutis was a friend of Humboldt, he died decades before the latter became aware of the dubious character of Keill's gossip.

While it was still unchallenged, Keill's gossip was brought, probably through the French translation of his book [12], to the notice of Jean-Sylvain Bailly (1736—1793), the great astronomer and revolutionary leader. Bailly explained that in both the Ptolemaic and Copernican systems virtually the same appearances would be presented by the biggest planets:

"But if the two smallest planets, Venus and Mercury, revolve around the earth while following the sun step by step, they must at all times appear full when they are beyond the sun [13]; almost always black and dark when they are on this side of the sun; and barely marked by a very thin crescent of light when they move to the right or the left of the sun. On the other hand, in Copernicus' system, in which they revolve around the sun, they should sometimes show a full disk, at other times a dark disk, and all the intermediate phases which we observe in the moon as it changes from its feeble crescent to its full and

---

[10] Robert S m i t h, *A Compleat System of Opticks.* Cambridge 1738, p. 415, section 1050.

[11] Guillermo Hernández de A l b a, *Crónica del Colegio Mayor de Nuestra Señora del Rosario.* Bogotá 1938—1940, vol. 2, p. 145; i d e m, *Copernico y los origines de nuestra independencia,* in: *Nicolás Copérnico.* Bogotá 1943, p. 19. The archival documents pertaining to the conflict between Mutis and the Dominicans were published in: John Tate L a n n i n g, *El Sistema de Copérnico en Bogotá.* "Revista de Historia de America", XVIII, 1944, 279—306.

[12] John K e i l l, *Institutions astronomiques.* Paris, 1746, pp. 267—268.

[13] But in the Ptolemaic system Venus and Mercury are never beyond the sun. Therefore they would never appear full, since they follow the sun step by step, and never come into opposition to it.

complete light [14]. Copernicus dared to proclaim that if our eye had the strength to see these two small planets as we see our satellite, we would perceive that they undergo the same variations" [15].

This story was told in 1779 by Bailly, as by Keill and Smith before him, without any indication of its source. But Jan Czyński (1801—1867) [16], a Polish refugee living in France, explicitly ascribed it to tradition, the favorite haunt of the mythopoeic mind: "Tradition has preserved for us some expressions used by Copernicus in defending his principles. In less enlightened times these expressions would have been taken for prophecies by a superior being. Some people maintained that his theory was false because phases of Venus and Mercury were not seen. «If Venus and Mercury», they said to him, «revolved around the sun and we revolved in a larger orbit, we ought to see them sometimes full and sometimes crescent-shaped, but that is what we never observe». «Nevertheless that is what happens», Copernicus replied, «and that is what you will see if you find a means of perfecting your sight»" [17].

This imaginary conversation between Copernicus and his opponents was published by Czyński in 1847, the very same year in which De Morgan demonstrated that the whole episode was the merest gossip. But De Morgan's demonstration appeared in a specialized periodical of limited circulation, whereas Czyński's book attracted many readers and exerted a powerful influence. Observe, for example, how the imaginary conversation was amplified in 1872 by the distinguished astronomer and highly successful popularizer of that science, Camille Flammarion (1842—1925):

"«If it were true», people said to Copernicus, «that the sun is at the center of the planetary system, and that Mercury and Venus revolve around the sun in orbits inside the earth's, these two planets should have phases. When Venus is on this side of the sun, it ought to be a crescent, like the moon when it sets in the evening; when Venus forms a right angle with the sun and us, it ought to present the aspect of the first quarter, and so on. Now that is what nobody has ever seen». «Nevertheless that is the fact», Copernicus answered, «and that is what

---

[14] Diagrams comparing the phases of Venus in the Ptolemaic and Copernican systems as well as in a low-power telescope were provided by Thomas S. K u h n, *The Copernican Revolution*. Cambridge, Mass. 1957; reprinted: New York 1959, p. 223; reviewed by Edward Rosen, see: "Scripta Mathematica", XXIV, 1959, pp. 330—331.

[15] Jean-Sylvain B a i l l y, *Histoire de l'astronomie moderne*. Paris 1779, vol. 2, p. 94.

[16] For a brief sketch of his life and character, see: Stanisław W ę d k i e w i c z, *Études coperniciennes*. Paris 1955—1957, pp. 58—61; reviewed by Edward Rosen, see: "Isis", L, 1959, pp. 177—178.

[17] Jan C z y ń s k i, *Kopernik et ses travaux*. Paris 1847, pp. 100—101; *cf.* pp. 16—17.

people will see some day if they find a means of perfecting their vision»" [18].

It is not surprising that Czyński's imagination prevailed over De Morgan's erudition in France. But even in England De Morgan was not effective enough to stop all the damage still being done by Keill of Oxford and Smith of Cambridge. For instance, thirty-five years after De Morgan had demolished the story, a "scholar of St. John's College, Cambridge" lauded Copernicus for making "the remarkable prediction that «if the sense of sight could ever be rendered sufficiently powerful, we should see phases in Mercury and Venus»" [19]. Toward the close of the nineteenth century another Englishman asserted: "Copernicus's conviction was so thorough that he predicted that we should see the phases of Mercury and Venus" [20].

At the turn of the century Edward Singleton Holden (1846—1914), former director of the Lick Observatory, told trustful American children that "Mercury and Venus show phases just as the moon does, and just as Copernicus had foretold that they would do... The prediction of Copernicus was correct; Venus had phases like the moon" [21]. A generation later adults in the United States were offered the same tale, embroidered with a little piety:

"As he [Copernicus] began to submit the outlines of his theory to some of his more intimate friends, another objection was brought to his attention. If Venus revolved about the sun, some argued, it should show phases like the moon, as its bulk, passing between earth and sun, obscured part of its light. The validity of this objection Copernicus quickly recognized, and asserted that here also one must wait upon the invention of more accurate instruments of observation. In God's good time, he added devoutly, the phases of Venus would be seen by human eyes. His prophecy was fulfilled in 1616 when Galileo's telescope showed them clearly" [22].

As the date for Galileo's discovery of the phases of Venus, 1616 is

---

[18] Camille F l a m m a r i o n, *Vie de Copernic*. Paris 1872, p. 207. Through the Spanish translation by Mariano Urrabieta: i d e m, *Vida de Copérnico*. Paris and Mexico City 1879, Flammarion's vivid version of the imaginary conversation was transmitted to: Oscar Miró Q u e s a d a, *Copérnico: su vida y su obra*. Lima 1950, pp. 110—111.

[19] Edward John Chalmers M o r t o n, *Heroes of Science: Astronomers*. London and New York 1882, p. 44. Since chapter 2 (pp. 32—62) is *On Copernik and His System*, Morton should have been included in: Henryk B a r a n o w s k i, *Bibliografia kopernikowska*. Warszawa 1958; reviewed by Edward Rosen, see: "Isis", XLIX, 1958, pp. 458—459.

[20] J. Villin M a r m e r y, *Progress of Science*. London 1895, p. 55.

[21] Edward S. H o l d e n, *Stories of the Great Astronomers*. New York 1900; reissued: New York and London 1912, pp. 110—111.

[22] Ernest R. T r a t t n e r, *Architects of Ideas*. New York 1938, pp. 25—26.

six years too late, and another recent popular work came closer to the mark: "At the beginning of 1611 Galileo published his discovery of the phases of Venus. According to the Copernican system these had to exist. But to the naked eye Venus had always appeared round. Lacking better proof, Copernicus had counted upon God's eventual help" [23].

Our last reverberator of the unfounded gossip started by Keill almost two and a half centuries ago is a historian and philosopher of science, Thomas S. Kuhn (b. 1922), who said: "Copernicus himself had noted in Chapter 10 of the First Book of the *De Revolutionibus* that the appearance of Venus could, if observable in detail, provide direct information about the shape of Venus's orbit" [24].

A curious feature of this most recent reverberation is that Kuhn's book [25] includes a translation of *De Revolutionibus*, I, 10, where Copernicus said nothing about the possibility of Venus' being "observable in detail". Neither there nor elsewhere did Copernicus make the statement "that the appearance of Venus could, if observable in detail, provide direct information about the shape of Venus's orbit". The shape of any planet's orbit, according to Copernicus, must be a circle or a combination of circles; as Kuhn himself translates [26] *De Revolutionibus*, I, 4, "the motion of the heavenly bodies is... circular... or composed of circular motions" [27].

By now we have seen ample evidence of the hardy persistence, despite De Morgan's valiant opposition, of Keill's gossip about "the prophecy of Copernicus" that future astronomers would discover the phases of Venus. The astronomer who did discover them was Galileo Galilei (1564—1642). He had converted to Copernicanism his former pupil and devoted friend Benedetto Castelli (1578—1643). On December 5, 1610, Castelli sent his beloved teacher a letter reading in part as follows:

"The Copernican system of the world is true, absolutely true, as I believe. Therefore Venus at equal distances from the sun must appear sometimes with horns and sometimes without horns, according as it will be on this side or on the other side of the sun. But such observations were impossible in previous centuries on account of the smallness of Venus' body and the disappearance of its form [when near the sun]. Now that with your immortal discoveries you have observed in the

---

[23] Herman K e s t e n, *Copernicus and His World*. New York 1945, London 1946, p. 369; in German: *Copernicus und seine Welt*. Amsterdam 1948, p. 443.

[24] T. S. K u h n, *op. cit.*, pp. 222—223.

[25] *Ibidem*, ed. 1957, pp. 176—179; ed. 1959, pp. 177—180.

[26] *Ibidem*, ed. 1957, p. 146; ed. 1959, p. 147.

[27] *Cf.*: *Nikolaus Kopernikus Gesamtausgabe*, ed. by Fritz Kubach, Franz Zeller and Karl Zeller. Munich and Berlin 1944—1949, vol. 2, p. 12, l. 25—26; also: p. 150, l. 22—23. Cited hereafter as *Gesamtausgabe*.

celestial domain so many other wonders invisible with ordinary powers, I should like to know whether you have made any observation in this regard, and whether what I have supposed is true" [28].

Castelli's supposition was confirmed by Galileo in the following reply to his favorite disciple on December 30, 1610: "About three months ago I began to observe Venus with the instrument, and I saw it round in shape and quite small [29]. From day to day it grew bigger while preserving the same round shape until finally, reaching quite a great distance from the sun, it started to lose its roundness on its eastern side, and in a few days it was reduced to a semicircle. It kept this shape for many days, while becoming larger in size. It is now commencing to become horned [30], and as long as it is visible in the evening, the horns will grow thinner until it disappears. But then when it returns in the morning, it will be seen with very thin horns turned away from the sun, and it will expand toward a semicircle up to its greatest distance [from the sun]. Then it will remain semicircular for some days, while decreasing in size. Afterwards it will pass from a semicircle to a full circle in a few days, and later it will be seen for many months as morning-star and evening-star, completely round but very small. The obvious consequences which follow herefrom are well known to you" [31].

These consequences were spelled out in detail in a communication which Galileo wrote on the same day, December 30, 1610, to the foremost contemporary Jesuit astronomer [32], who was not a Copernican: "Venus (and Mercury unquestionably does the same thing) goes around the sun,

---

[28] *Le Opere di Galileo Galilei, Edizione Nazionale.* Vol. 1—20. Firenze 1890—1909; reprinted: 1929—1939. Cited hereafter as *EN*. Here: *EN*, vol. 10. p. 482, l. 13—21. The circumstances surrounding this letter were distorted by a priest intent on besmirching Galileo's reputation. For a crushing rebuttal of the priest, who had not been appointed to the editorial commission of the national edition of Galileo's works, see: Antonio F a v a r o, *Galileo Galilei, Benedetto Castelli e la scoperta delle fasi di Venere.* "Archivio di Storia della Scienza", I, 1919—1920, pp. 284—293; cf. also: pp. 276—277.

[29] Since Galileo says that he began to observe Venus with the telescope about the end of September 1610, Humboldt's statement that Galileo saw Venus crescent-shaped in February 1610, misdated the discovery of the phases of Venus by more than half a year. *Cf.*: A. v. H u m b o l d t, *op. cit.*, vol. 2, p. 362.

[30] This emphasis on the gradualness of the changes in Venus' appearance (*di giorno in giorno, finalmente, in pochi giorni, molti giorni, hora*) demonstrates the erroneousness of the undocumented reference to "Galileo's statement that Venus altered its appearance by leaps and bounds (*sprunghaft*)" in: Ernst Z i n n e r, *Entstehung und Ausbreitung der coppernicanischen Lehre.* Erlangen 1943, p. 343; reviewed by Edward Rosen, see: "Isis", XXXVI, 1945—1946, pp. 261—266.

[31] *EN*, vol. 10, p. 503, l. 16—31.

[32] Christopher Clavius (1538—1612), who was the subject of a useful little article by Otto Meyer; see: O. M e y e r, *Christoph Clavius Bambergensis.* "Kleine Veröffentlichungen der Remeis-Sternwarte Bamberg", XXXIV, 1962, pp. 137—143.

which is without any doubt the center of the principal revolutions of all the planets. Moreover, we are certain that the planets are in themselves dark, and shine only when illuminated by the sun. That this does not happen with the fixed stars, I believe as a result of some of my observations, and that the [actual] planetary system is surely different from the one which is generally accepted" [33].

In his magnificent *Dialogue* (*Dialogo sopra i due massimi sistemi del mondo, tolemaico e copernicano*, Florence, 1632), which was condemned as heretical by the Roman Catholic church, Galileo explained to his readers, who were not assumed to be professional astronomers, how the phases of Venus prove that this planet rotates around the sun:

"It never moves further away from the sun than a certain definite interval of some 40°, so that it never becomes opposite the sun, nor at right angles to it, nor even at an angle of 60° to it. Moreover, it appears almost 40 times larger at one time than at another, being very big when it proceeds in the retrograde direction toward its evening conjunction with the sun, and very small when it moves in the forward direction towards its morning conjunction. In addition, when it appears biggest, it shows a horned shape; when it appears smallest, it is seen perfectly round. Since, I say, these phenomena are true, I do not see how it is possible to avoid the statement that this planet revolves in a circle around the sun. This circle can nowise be said to embrace and contain the earth within itself, nor to be below the sun (that is, between the sun and the earth), nor to be above the sun. This circle cannot embrace the earth, because [in that case] Venus would sometimes become opposite the sun. Nor can the circle be below the sun, because Venus would appear horned near both its conjunctions with the sun. Nor can the circle be above the sun, because the planet would always look round and never horned" [34].

Although Galileo's discovery of the phases of Venus proved Copernicus' thesis that the planet revolves around the sun, Copernicus himself never actually saw Venus' phases. They are not visible to the naked eye, and the telescope was not invented until nearly half a century after Copernicus died in 1543. In Galileo's *Dialogue,* when one of the interlocutors asks why the phases of Venus were concealed from Copernicus and were revealed later, the principal spokesman for Galileo answers as follows:

"These things can be grasped only with the sense of sight, which nature did not give to mankind so perfect that it could succeed in discerning such differences. Indeed the organ of sight makes trouble for itself. But in our age it pleased God to grant to human ingenuity an invention

---

[33] *EN*, vol. 10, p. 500, l. 36—42.
[34] *EN*, vol. 7, pp. 351, l. 21—352, l. 1.

so remarkable that it could improve our vision by increasing it 4, 6, 10, 20, 30, and 40 times. Thereafter countless objects which had been invisible to us either on account of their distance or on account of their extremely small size were made perfectly plain by means of the telescope" [35].

The spokesman then goes on to say that when Venus approaches its evening conjunction, "the telescope clearly shows us its horns just as definite and well-marked as those of the moon. The horns of Venus look like part of a very big circle, and they are almost 40 times larger than the disk of Venus when it is above the sun and making its last morning appearance". Thereupon the questioner exclaims: "O Nicholas Copernicus, what joy would have been yours to see this part of your system confirmed by such clear observations!" [36].

Do we not have in the foregoing passages of Galileo's *Dialogue* some of the raw materials out of which Keill built his unhistorical gossip about "the prophecy of Copernicus"? To locate the rest of Keill's raw materials, let us now look at what Copernicus said about the phases of Venus. In his *Revolutions (De Revolutionibus orbium coelestium*, Nuremberg, 1543), I, 10, he discussed the arrangement of the planets in space:

"With regard to Venus and Mercury differences of opinion are found, because these planets do not pass through every angular distance from the sun, as the other planets do. Therefore some people, like Timaeus in Plato, locate Venus and Mercury above the sun. Other people, like Ptolemy and a good many of the recent writers, place Venus and Mercury below the sun. Al-Bitruji puts Venus above the sun [37] and Mercury below it. Plato's followers believe that all the heavenly bodies, being otherwise dark, shine because they receive the light of the sun. Hence, if Venus and Mercury were below the sun, since their angular distance from it is not very great, they would look semicircular or at any rate less than completely round. For, the light which they receive would be reflected mostly upward, that is, toward the sun, as we see in the new or the dying moon" [38].

The foregoing passage clearly shows us that the phases of Venus did not have the same significance for Copernicus as they subsequently acquired. For they were interpreted by Galileo, the first human being who ever saw them, as confirmation of Copernicus' contention that Venus revolves around the sun and not around the earth. But in

---

[35] *EN, ibidem*, p. 363, l. 15—23.

[36] *EN, ibidem*, p. 367, l. 5—12.

[37] This statement has been turned topsy-turvy by some recent writers, whose errors were corrected in: Edward R o s e n, *Copernicus and Al-Bitruji.* "Centaurus", VII, 1961, pp. 152—156.

[38] *Gesamtausgabe*, vol. 2, p. 22, l. 10—19.

Copernicus' thinking the phases of Venus belong in a completely different context. They have nothing to do with the debate whether Venus revolves around the sun or around the earth. They are discussed entirely within the framework of the geocentric theory.

Assume that the earth is motionless at the center of the universe. The moon is the earth's nearest neighbor, and it revolves around the earth. Which of the other heavenly bodies revolving around the earth is its second nearest neighbor? "The sun", answered Plato [39]. In defense of Plato's arrangement (central earth, moon, sun, then Venus and Mercury), and in opposition to a rival geocentric arrangement (earth, moon, Mercury, Venus, sun), the following argument was developed by Platonists [40]. If you put Venus between the earth and the sun, then Venus should show phases like those of the moon, which is also a dark body between the earth and the sun. But Venus shows no such phases to the unaided eye, which sees this planet perfectly round whenever it is visible. This absence of the phases of Venus, argued the Platonists, supports Plato's version of the geocentric theory as against Ptolemy's version of the geocentric theory.

In the Ptolemaic system, Venus should show phases. Their non-appearance was regarded by Copernicus as a defect in the Ptolemaic system, which enjoyed virtually universal support when he was writing his *Revolutions* in the sixteenth century. But two hundred years later, in Keill's time, the Ptolemaic system had long been dead, having suffered fatal blows at the hands of Copernicus, Galileo, Kepler, and Newton. Under these altered conditions the absence of Venus' phases was anachronistically transformed by Keill from an anti-Ptolemaic argument (as it had been for Copernicus) into an anti-Copernican argument as a basis for his mythical "prophecy of Copernicus".

If we have correctly identified the passages in Galileo's *Dialogue* and Copernicus' *Revolutions* which served Keill as the ingredients from which he concocted his "prophecy of Copernicus", we now face a more difficult question: how did Copernicus explain the absence of Venus' phases? Recalling the Platonists' use of this argument against Ptolemy was no doubt an effective maneuver on Copernicus' part. But what about Venus' phases in Copernicus' own system? Of course he never saw them. Nor, despite Keill, Smith, Mutis, Bailly, Czyński, Flammarion

---

[39] Plato, *Timaeus*, 38 C—D.

[40] Not by Plato himself, as in the English translation of the Preface and Book I of the *De Revolutionibus* by John F. Dobson and Selig Brodetsky, "Occasional Notes of the Royal Astronomical Society", II, 1947, p. 16; reissued: 1955; and: T. S. Kuhn, *op. cit.*, ed. 1957, p. 176; ed. 1959, p. 177; and: Milton K. Munitz, *Theories of the Universe.* Glencoe 1957, London 1958, p. 165. Copernicus says: "...those who follow Plato" ("...*qui Platonem sequuntur*"), *cf.*: *Gesamtausgabe*, vol. 2, p. 22, l. 15.

and their followers, did Copernicus, who was not a clairvoyant, foresee
that Venus would some day be found to have phases.

How, then, did Copernicus explain their non-appearance in his own
pre-telescopic times? To this question Humboldt gave the right answer
by pointing out that Copernicus discusses "the doubts which the more
modern adherents of the Platonic opinions advance against the Ptolemaic system on account of the phases of Venus. But in the development
of his own system Copernicus does not speak explicitly about these
phases" [41]. As Galileo's spokesman in the *Dialogue* remarks about another serious problem in the Copernican system, "Maybe Copernicus
himself could not find a solution of it which satisfied him completely,
and perhaps for that reason he kept quiet about it" [42].

The Copernican system, an immense revolution in human thought,
brought in its train a host of perplexing difficulties. Some of them were
solved correctly by Copernicus, but some of the solutions proposed by
him have turned out to be wrong. About other questions he remained
silent, as Galileo's spokesman put it, "because he could not explain to
his own satisfaction a phenomenon so contrary to his system. And yet,
convinced by so many other indications, he stuck to his theory and held
it to be true... These are the difficulties which make me wonder about
Aristarchus and Copernicus. They must have noticed them, and then
could not solve them. Yet, as a result of other remarkable confirma-
tions, they trusted so much in what reason told them that they confidently asserted that the structure of the universe could have no other
form than the one described by them" [43].

But with regard to the phases of Venus Galileo did not believe that
Copernicus resorted to the strategy of silence. In his *Sunspots (Istoria
e dimostrazioni intorno alle macchie solari*, Rome, 1613), Galileo remarks
that the anti-Copernicans will explain the absence of phases in Venus
by saying that "either Venus is self-luminous or its substance is pene-
trable by the sun's rays, so that it is illuminated not only on its surface
but also throughout its entire depth". The anti-Copernicans "can have
the courage to shield themselves with this reply because there has been
no lack of philosophers and mathematicians who held this belief... Copernicus himself has to accept one of the aforementioned theories as possible,
or rather as necessary, since he could not explain why Venus does not
look horned when it is below the sun. In fact nothing else could be said
before the coming of the telescope let us see that Venus is actually as
dark as the moon and that, like the moon, it changes its shape" [44]. This

[41] A. v. H u m b o l d t, *op. cit.*, vol. 2, p. 362.

[42] *EN*, vol. 7, p. 194, l. 3—5.

[43] *EN, ibidem*, pp. 362, l. 13—363, l. 1.

[44] *EN*, vol. 5, pp. 99, l. 16—100, l. 1.

was what Galileo wrote on May 4, 1612, about Copernicus' attitude toward the phases of Venus.

Nearly two decades later in the *Dialogue* Galileo reiterated the same view: "If the body of Venus is itself dark and, like the moon, shines only because it is illuminated by the sun, as seems reasonable, when Venus is below the sun it ought to look horned, like the moon when it is similarly near the sun. The phenomenon is not visible in Venus. Copernicus therefore declared that it [45] was either self-luminous or made of such material that it could imbibe sunlight and transmit it throughout its entire depth, so that it could always look bright to us. In this way Copernicus accounted for the absence of phases in Venus" [46].

Actually Copernicus did not account for the absence of phases in Venus in this way or in any other way. He himself expressed no opinion about the matter, as Humboldt correctly said. In the *Revolutions,* I, 10, we recall, Copernicus has the followers of Plato agree with the followers of Ptolemy in placing the earth at the center of the universe. But these two schools disagree about the position of Venus, the Platonists putting it above the sun, and the Ptolemaists below the sun. Copernicus then proceeds to say, as we saw above:

"Plato's followers believe that all the heavenly bodies, being otherwise dark, shine because they receive the light of the sun. Hence, if Venus and Mercury were below the sun, since their angular distance from it is not very great, they would look semicircular or at any rate less than completely round. For, the light which they receive would be reflected mostly upward, that is, toward the sun, as we see in the new or the dying moon".

This objection by the Platonists to the Ptolemaic theory, Copernicus reports, was answered by the Ptolemaists, who "say [*fatentur*] that in the planets there is no opacity like the moon's. On the contrary, these bodies shine either with their own light or with the sunlight absorbed throughout their bodies" [47].

Although these ideas about the nature of the planets were ascribed to the Ptolemaists by Copernicus, Galileo attributed them to Copernicus himself. He did so because the first edition of the *Revolutions* (Nuremberg, 1543) put the verb "say" in the first person (*fatemur*) [48]. This

---

[45] Venus, not the moon, as in: Giorgio de S a n t i l l a n a, *Galileo Galilei, Dialogue on the Great World Systems.* Chicago 1953, p. 343. This error was introduced by Santillana in his revision of the translation of Galileo's *Dialogue* by Thomas Salusbury in *Mathematical Collections and Translations,* London 1661—1665, tome 1, part 1, p. 302; see: Stillman D r a k e, *A Kind Word for Salusbury.* "Isis", XLIX, 1958, p. 27.

[46] *EN,* vol. 7, p. 362, l. 3—12.

[47] *Gesamtausgabe,* vol. 2, pp. 22, l. 35—23, l. 1.

[48] Fol. 8r, line 13.

typographical error was corrected in the second edition of the *Revolutions* (Basel, 1566), which shifted the verb "say" to the third person (*fatentur*) [49]. Copernicus' holograph manuscript [50] plainly shows that the reading in the first edition was wrong: it is not "we" (the author), but "they" (the Ptolemaists), who say that the planets are either self-luminous or transparent.

Galileo had a copy of both the first and the second edition of the *Revolutions* [51], but unfortunately in this matter he relied on the first edition. I say "unfortunately", because Galileo's magisterial prestige induced others to accept his statement that it was Copernicus who explained the absence of the phases of Venus by describing the planet as either self-luminous or transparent. Thus, according to a note in the 1744 edition of Galileo's works, "Copernicus wrote that either Venus was self-luminous or it absorbed sunlight throughout its entire depth so that it could appear bright even when it turns and shows us the part of its globe that the sun does not strike" [52].

Of the two alternatives supposedly adopted by Copernicus, the self-luminosity of Venus was dropped by a biographer of Galileo, John Elliot Drinkwater Bethune (1801—1851), who retained only the transparency: "Copernicus, whose want of instruments had prevented him from observing the horned appearance of Venus when between the earth and sun, had perceived how formidable an obstacle the non-appearance of this phenomenon presented to his system; he endeavoured, though unsatisfactorily, to account for it by supposing that the rays of the sun passed freely through the body of the planet" [53].

The second alternative was reinstated by De Morgan, when reporting on this subject to the Royal Astronomical Society in the address from which we have already read an excerpt: "If we try to examine what the opinion of Copernicus on this matter really was, a point of some little curiosity arises. It depends on one word, whether he did or did not assert his belief in one or other of these two opinions — that the

---

[49] *Loc. cit.*

[50] *Gesamtausgabe*, vol. 1, fol. 8r, line 3 up.

[51] *Cf.*: Antonio F a v a r o, *La libreria di Galileo Galilei*. "Bullettino di Bibliografia e di Storia delle Scienze Matematiche e Fisiche" (Boncompagni), XIX, 1886, pp. 246—247; the entire Boncompagni's "Bullettino" has just been re-issued by Johnson Reprint Corporation of New York and London. For Galileo's autograph notes on Copernicus' *Revolutions* see: A. F a v a r o, *Nuovi studi galileiani*. Venice 1891, pp. 76—78 (*Postille galileiane all' opera capitale di Niccolò Coppernico*).

[52] *Opere di Galileo Galilei*. Padua 1744, vol. 2, p. 36; presumably this unsigned note was written by the editor, Giuseppe Toaldo (1719—1798).

[53] J. E. D. B e t h u n e, *Life of Galileo*. London 1833, p. 35; New York 1835, p. 31. This is not the only error committed by Bethune, "whose scholarship and minute accuracy are beyond question", in the excessively generous judgment of S. Drake. see: "Isis", XLIX, 1958, p. 29.

planets shine by their own light, or that they are saturated by the solar light, which, as it were, soaks through them. I support the affirmative: that is to say, I hold it sufficiently certain that Copernicus did express himself to the effect that one or the other of these suppositions was the truth" [54].

De Morgan's restoration of the second alternative was overlooked by John Joseph Fahie (1846—1934), a close friend of the editor of the national edition of Galileo's works. Despite such guidance Fahie's biography of Galileo echoed Bethune's twofold error: "Copernicus himself had endeavoured to account for this [absence of Venus' phases], by supposing that the sun's rays passed freely through the body of the planets" [55].

The same double mistake was made by John Gerard (1840—1912), provincial of the Jesuits in England and author of the article on Galileo in the *Catholic Encyclopedia*: "It had been argued against the said system [of Copernicus] that, if it were true, the inferior planets, Venus and Mercury, between the earth and the sun, should in the course of their revolution exhibit phases like those of the moon, and, these being invisible to the naked eye, Copernicus had to advance the quite erroneous explanation that these planets were transparent and the sun's rays passed through them" [56].

That Venus was transparent or self-luminous was a theory imputed by Copernicus to the followers of Ptolemy. It was not Copernicus' own conception of the physical nature of the planet, despite Galileo, Toaldo, Bethune, De Morgan, Fahie, and Father Gerard. All but one of these writers merely enunciated the unsupported dictum that Copernicus believed Venus to be transparent or self-luminous. The single exception is De Morgan, whose attempt to justify this statement about Copernicus' belief will be examined in a moment. The others simply copied from Galileo, or from those who had previously copied from Galileo. But, as we saw above, Galileo was misled by a misprint, which he would doubtless have detected had he compared the text of the first edition of the *Revolutions* with the text of the second edition. We know from his letter of August 19, 1610, to Kepler that he despised the comparers of texts for trying to learn the truth about nature from books [57]. Of course, if we want to know nature, we must read the Book of Nature,

---

[54] "Monthly Notices of the Royal Astronomical Society", VII, 1847, p. 291.

[55] John Joseph F a h i e, *Galileo, His Life and Work*. London 1903; reprinted: Dubuque 1962, p. 124; i d e m, *The Scientific Works of Galileo*, in: *Studies in the History and Method of Science*, ed. by Charles Singer. Oxford 1917—1921, vol. 2, p. 239.

[56] *Catholic Encyclopedia*. New York 1907—1914, vol. 6, p. 343.

[57] *EN*, vol. 10, p. 423, l. 59—62; Johannes K e p l e r, *Gesammelte Werke*. Vol. 16. Munich 1954, p. 329, l. 58—61.

that is, the physical world. But if we want to know what Copernicus thought about Venus, the Book of Nature cannot help us. We must read the book of Copernicus; and if its first edition contains a crucial misprint, comparison of texts will help us to detect that fact.

Such a comparison was instituted by De Morgan. But despite his enormous erudition, in this instance he went astray [58]. Concerning the Ptolemaic arrangement of Mercury and Venus between the earth at the center and the sun, De Morgan said that Copernicus "describes the opinion just mentioned favourably, referring, not to his own view, but to that of those others who had held it. This is not an uncommon idiom: persons advocating an unpopular opinion are very apt to describe the maintainers of it in the third person, though themselves be of the number... Copernicus is evidently speaking with approbation of the opinions which he describes; and it would be difficult to say why *comperiunt* or *putant* in one sentence should imply approbation, and *fatentur,* in the next, should be at least disavowal, if not disapprobation" [59].

De Morgan evidently made the foregoing analysis with less than his customary care, as is indicated by his misquotation of *putant* instead of *supputant,* the word actually used by Copernicus [60]. Much more serious are De Morgan's two mistakes about the Copernicus passage under consideration. In the first place, the Ptolemaic arrangement of the planets was not "an unpopular opinion" in Copernicus' time; in fact he says, as we saw above, that it was held by "a good many of the recent writers" (*bona pars recentiorum*). Did De Morgan anachronistically transpose the nineteenth-century unpopularity of the Ptolemaic planetary arrangement back to the sixteenth century?

However this may be, De Morgan committed a second error in saying that Copernicus "describes the opinion just mentioned favourably", and "is evidently speaking with approbation of the opinions which he describes". Actually Copernicus describes the opinion neither with approbation nor with disapprobation, neither favorably nor unfavorably. He describes it dispassionately and accurately. Then he proceeds to present powerful arguments against it. He is "referring, not to his own view, but to that of those others who had held it", as De Morgan correctly saw. The view in question is the Ptolemaic planetary arrangement. To disprove this view was Copernicus' prime purpose in writing the *Revolutions*. In sum, then, De Morgan's attempt to show that Copernicus

---

[58] For other examples, see: Edward R o s e n, *De Morgan's Incorrect Description of Maurolico's Books.* "Papers of the Bibliographical Society of America", LI, 1957, pp. 111—118; i d e m, *Maurolico's Attitude toward Copernicus.* "Proceedings of the American Philosophical Society", CI, 1957, pp. 177—194.

[59] "Monthly Notices of the Royal Astronomical Society", VII, 1847, pp. 291—292.

[60] *Gesamtausgabe,* vol. 1, fol. 8r, line 4 up; *op. cit.,* vol. 2, p. 22, l. 33.

conceived Venus to be self-luminous or transparent must be adjudged a complete failure. Copernicus himself did not regard Venus as self-luminous or transparent. He ascribed that opinion to the followers of Ptolemy.

Since Copernicus did not hold Venus to be self-luminous or transparent, what did he think was the source of the planet's light? Did he, like "Plato's followers, believe that all the heavenly bodies, being otherwise dark, shine because they receive the light of the sun"? Just as he did with the followers of Ptolemy, Copernicus describes this opinion of the Platonists dispassionately. He neither approves it nor disapproves it. Having no decisive evidence for or against the Platonists' view, Copernicus does not say how Venus obtains its light. Refraining altogether from treating this question, he leaves it for others to decide.

Since we do not know from Copernicus' own statements what he believed the source of Venus' light to be, can we perhaps make an inference from the prevailing contemporary or traditional opinion about the subject? In other words, can we put our trust in Galileo's deadliest opponent? Finding no shadow cast by Venus on the sun in a predicted conjunction of those two bodies, the Jesuit Christopher Scheiner (1573—1650), or Apelles, as he then called himself, wished to dispose of the possible explanation that:

"The planet Venus does not produce a shadow or spot for us because it is endowed with its own light, which is not, like the moon's, received from the sun. But this will be contradicted by experience, reason, and the common agreement of all the ancient and modern mathematicians" [61].

Is it true, as Scheiner-Apelles maintained, that "all the ancient and modern mathematicians" agreed in denying the self-luminosity of Venus? Or was Galileo right in insisting that, according to some philosophers and mathematicians, Venus has its own light "and let this be said by leave of Apelles, who writes otherwise" [62]. As one example among many, Galileo could have pointed to Al-Bitruji, who declared:

"What convinces me that Mercury and Venus do not receive their light from the sun nor from outside themselves is the fact that we always see them shining when they are near the sun... If their light were, like the moon's [derived from the sun], the bright part of Mercury would always be crescent-shaped because its angular distance from the sun is not very great, and the same is true for Venus" [63].

[61] Christopher S c h e i n e r, *Tres epistolae de maculis solaribus.* Augsburg 1612, fol. A4v; reprinted: *EN*, vol. 5, p. 28, l. 26—29; *cf.*: Ch. S c h e i n e r, *Accuratior disquisitio.* Augsburg 1612, p. 14; reprinted: *EN, ibidem*, p. 46, l. 17—18.

[62] *EN, ibidem*, p. 99, l. 21; *cf.*: p. 197, l. 19—24.

[63] A l - B i t r u j i, *De motibus coelorum*, ch. 16, no. 11; i d e m, *ibidem*, ed. by Francis J. Carmody. Los Angeles 1952, p. 128.

Whereas Al-Bitruji believed in the self-luminosity of Venus, its absorption of sunlight was defended by Regiomontanus (1436—1476), the greatest astronomer of the fifteenth century: "The bodies of the planets other than the moon absorb sunlight into themselves. They do so to no greater extent than the moon. Yet [64], perhaps on account of the different variation in the planets and stars, the planets other than the moon receive the sun's rays into their very depths. On the other hand, on account of its greater density, the moon is not illuminated down to its center. Hence it looks to us like a crescent. But Venus, even though it is quite close to the sun, never appears in this way as a crescent, because its body is penetrated throughout by sunlight" [65].

We need look no further than Al-Bitruji's belief in the self-luminosity of Venus and Regiomontanus' conception of that planet's thorough absorption of sunlight to decide between Galileo and Scheiner. The great Italian was right, and the Jesuit was wrong in contending that by "the common agreement of all the ancient and modern mathematicians" Venus was a dark body. There was no such common agreement. There was in fact sharp disagreement. On both sides of the question Copernicus had ample and respectable authority. But he had no observational evidence to decide between the opposing opinions. He did not foresee the invention of the telescope. Nor did he ever predict that that marvelous instrument would some day disclose the phases of Venus. With regard to the source of that planet's light, his prudent silence anticipated the wise counsel of the eminent twentieth-century philosopher who said: *"Wovon man nicht sprechen kann, darüber muss man schweigen"* [66].

---

[64] Reading *tamen* rather than *tantum*. Much more serious blunders by F. J. Carmody were corrected in: Edward R o s e n, *Regiomontanus' Breviarium.* "Medievalia et Humanistica", XV, 1963, pp. 95—96.

[65] Francis J. C a r m o d y, *Regiomontanus' Notes on Al-Bitruji's Astronomy.* "Isis", XLII, 1951, p. 129, no. 32—33.

[66] Ludwig W i t t g e n s t e i n, *Tractatus logico-philosophicus.* Reprinted: London 1961, New York 1961, pp. 150—151.

11

Copernicus' Axioms

The earliest substantial scientific treatise which correctly viewed the earth as a moving planetary satellite of the sun is the *Commentariolus* of Nicholas Copernicus (1473–1543). This work begins with a succinct summary of the two principal anterior astronomical systems, both based on a motionless earth located in the center of the universe. The irremediable defects of these two ancient theories (Eudoxus' homocentrics and Ptolemy's eccentric epicycles) are then reviewed by Copernicus for the purpose of explaining to his readers why he has undertaken to paint his new picture of the cosmos. He is prepared to tackle this unprecedented task "if some postulates, which are called axioms, are granted to me" (*si nobis aliquae petitiones, quas axiomata vocant, concedantur*). There follow at once seven "postulates, which are called axioms."

A recent translator of the *Commentariolus* charges that these seven postulates are "incorrectly called *axioms* since they are hardly self-evident."[1] For example, Postulate 2 begins: "The center of the earth is not the center of the universe." Far from being self-evident, the displacement of the earth from the center of the universe had been vigorously assailed by Aristotle and Ptolemy, among others, and rejected by nearly every one before Copernicus. Yet he says that his postulates, none of which is self-evident (in fact, they startled and shocked his contemporaries), "are called axioms." Didn't Copernicus, who studied mathematics and astronomy for many years at Polish and Italian universities, where the language of instruction was Latin, know the meaning of "axiom"?

This term was avoided by Euclid, whose geometry provides the earliest extant model of the axiomatic method. The nature of an axiom was extensively discussed, however, centuries later by Proclus in his *Commentary on the First Book of Euclid's Elements*. According to this influential commentator, "the axioms embrace what is of itself at once evident to the mind and readily grasped by our uninstructed intelligence";

"the axiom asserts something essential and immediately acknowledged by the auditors, for instance, that fire is hot";

"the foregoing are what are universally called the indemonstrable axioms, since everybody thinks that they are true and no one disputes them";

"some people ... bestow the name 'axiom' on a proposition that is immediate and self-evident by reason of its clarity";

"all the axioms must be presented as immediate and self-evident, known and credible of themselves."[2]

Despite his own insistence on the self-evident character of axioms, Proclus was uncomfortably aware that the great Archimedes had begun his *Equilibrium of Planes* by "postulating ... whereas one might rather label this an axiom."[3] Proclus also lamented that "the Stoics are in the habit of calling every simple declarative sentence an axiom."[4] A Stoic axiom, however, did not even need to be affirmative, according to Diogenes Laertius' account of their teachings:

"An axiom is that which is either true or false .... An axiom is the affirmation or denial of an independent statement, such as 'It is day,' 'Dio is walking around'."[5]

The Stoic conception of an axiom, it is clear, was not followed by Copernicus, whose "postulates, which are called axioms," were true, not false, in his opinion. Nor did he share Proclus' attitude toward the axiom as a self-evident proposition of the type "fire is hot." For, Copernicus asked his readers to grant (*concedantur*) his "postulates, which are called axioms," whereas Proclus' axioms imposed themselves on every reasonable person, who did not have to be asked to grant anything.

What, then, are Copernicus' "postulates, which are called axioms"? As a group, they are the propositions underlying a train of consecutive demonstrations, as defined in Aristotle's *Posterior Analytics*:

"An axiom is a primary proposition which must be possessed by whoever is to gain any knowledge";

"the axioms are the primary propositions from which a proof proceeds."[6]

To this Aristotelian type, then, Copernicus' axioms conform. They are not false assertions, as the Stoic axioms might be; nor are they self-evident truths, like Proclus' axioms. They are underlying propositions from which an extended series of demonstrations starts. This is what Copernicus meant when he said that his postulates "are called axioms." In using this terminology, he was of course absolutely correct. When his recent translator asserts that Copernicus' postulates are "incorrectly called axioms," it is the translator, not Copernicus, who is incorrect.

Not only were these postulates "incorrectly called axioms," according to the translator, Professor Noel M. Swerdlow of the University of Chicago, but "the introduction of these postulates at this point appears unmotivated" to Swerdlow.[1] His failure to understand Copernicus' motive in placing the postulates where they are is connected with Swerdlow's misapprehension of their essential nature. They follow Copernicus' rejection of the ancient theories, and precede the exposition of his own system. In short, Copernicus introduced these postulates exactly where they logically belong. Although no other place would have been suitable for them, and no better place is suggested by Swerdlow, he condemns Copernicus' indispensable arrangement as "this flaw," which is "intelligible if one considers that the *Commentariolus* may well have been written in haste with no revision."[1] Actually, in its admirable compactness, without a superfluous word, the *Commentariolus* gives every sign of being the end product of much reflection, careful planning, and superb organization.

Swerdlow, however, detects "the logical error of stating postulates 2, 4, 5, and 7 as postulates rather than deductions from postulates 3 and 6."[1] Treating deducible propositions as postulates would have been a logical error had Copernicus been engaged, like some nineteenth- or twentieth-century axiomatist, in devising a purely formal, abstract, logico-mathematical, hypothetico-deductive system, devoid of specific existential content. Instead, Copernicus was actually composing an epoch-making treatise which portrayed the physical universe in an unconventional way. Unlike Giuseppe Peano nearly four centuries later, Copernicus was not in the least concerned "to enunciate as postulates only those that cannot be deduced from others which are simpler" and "to reduce them to the smallest number."[7] These requirements (that every axiom must be logically independent of all its fellow-axioms, which must be minimized in number) were canonized as part of the fairly recent campaign to make the foundations of mathematics absolutely rigorous. However, to retroject the stern ambitions of modern axiomatics onto Copernicus, and to impute to him a "logical error" explicated centuries after his death, is of course grossly anachronistic. Awareness of the chronological development of the various disciplines is a distinguishing characteristic of serious history of science, which Swerdlow apparently aspires to write.

In composing the *Commentariolus*, Copernicus did not endeavour to anticipate Peano's *Sui fondamenti della geometria* or David Hilbert's *Grundlagen der Geometrie*.[8] Instead, for his structural model Copernicus looked back toward some such earlier brief discussion as *De insidentibus in humidum*, whose author was then believed to have been none other than the incomparable Archimedes himself. This medieval pseudo-Archimedean treatise *On Floating Bodies* labeled its six primary propositions *petitiones*, the very word used by Copernicus in the *Commentariolus* for his seven "postulates, which are called axioms" (*petitiones, quas axiomata vocant*). Like the *petitiones* in Copernicus' *Commentariolus*, pseudo-Archimedes' *petitiones* are intended as true statements about the real world. They rise in an unmistakable crescendo from

(1) "No body is heavy in relation to itself," and
(2) "Every body weighs more in air than in water," to
(6) "If equal [volumes of] bodies are equal in weight, their specific gravities are said to be equal."[9]

By the same token, in the *Commentariolus* Copernicus' seven *petitiones* mount in a magnificent ascent. (1) undermines Eudoxus' principle of homocentricity by denying that everything in the universe is centered on the earth: "There is no one center." Consequently there are multiple centers, of which the earth is one. (2) puts the earth at the center, not of the universe, but of the moon's motions. Another center is the sun, placed by (3) near the center of the universe. The distance between these two centers, earth and sun, is said by (4) to be imperceptible in comparison with their distance from the stars. (5) The stars do not move, their apparent daily motion being due to the earth's axial rotation. (6) The sun does not move, its apparent motion being due to the earth's orbital revolution. (7) The apparent retrogression of the planets is likewise only an appearance due to the motion of the terrestrial observer. By common consent, (7) is Copernicus' single most valuable contribution to technical astronomy. It is concluded by his triumphant jubilation:

> "The motion of the earth alone, therefore, suffices to explain so many apparent irregularities in the heavens."

With regard to (7), the first correct explanation, in mankind's entire history, of the planets' apparent loops, Swerdlow pontificates: "since it can be *proved* from postulates 3 and 6, it should not be stated as a postulate."[10] This belated advice is based on Swerdlow's misconception of the role of the postulates in Copernicus' *Commentariolus*. Moreover, according to Swerdlow, "Copernicus has raised no objection to Ptolemy's representation of the second anomaly,"[1] that is, the planetary loops. But these are reduced by (7) to the rank of mere appearances or optical illusions. Can there be any more forceful "objection to Ptolemy's representation of the second anomaly"?

Also with regard to (6), "What appear to us as motions of the sun are due, not to its motion but to the motion of the earth," Swerdlow has some tardy advice. He feels that "the apparent diurnal motion of the sun ... is really unnecessary here and should have been attached to postulate 5."[10] But the subject of (5) is the stars, not the sun.

In astronomical treatises earlier than Copernicus' *Commentariolus*

"it is shown that the earth is but a point in comparison to the distance of the sphere of the fixed stars .... Now it is also pointed out [elsewhere than in the *Commentariolus*] that for all practical purposes the earth is also a point compared to the distance of the sun, and so [according to Swerdlow] Copernicus begins [(4)] with this comparison."[10]

This comparison does not begin (4), which reads as follows:

"The ratio of the earth's distance from the sun to the height of the firmament is so much smaller than the ratio of the earth's radius to its distance from the sun that the distance between the earth and the sun is imperceptible in comparison with the height of the firmament."

Copernicus begins (5) as follows:

"Whatever motion appears in the firmament is due, not to it, but to the earth."

Then he adds:

"The earth ... performs a complete rotation on its fixed poles in a daily motion, while the firmament and highest heaven abide unchanged" (*Terra ... motu diurno tota convertitur in polis suis invariabilibus firmamento immobili permanente ac ultimo caelo*).

The last six words are translated by Swerdlow as follows:

"the sphere of the fixed stars remains immovable and the outermost heaven."[11]

But Copernicus did not mean that the sphere of the stars remains the outermost heaven. For Copernicus, the starry sphere, which is the outermost heaven, remains immovable, its apparent daily rotation being due to the real axial rotation of the earth.

Swerdlow tries to support his (mis)translation by remarking: "The sphere of the fixed stars is the outermost heaven because a sphere of diurnal rotation is no longer necessary."[10] But a sphere of diurnal rotation is still necessary, since diurnal rotation is an astronomical fact, and for Copernicus that necessary sphere of diurnal rotation is the earth. For Ptolemy, the diurnal rotation had been performed by the starry sphere, which was his outermost heaven. Hence, Ptolemy's outermost heaven moved. In order to emphasize his divergence in this respect from Ptolemy, Copernicus said in (5) "the firmament and highest heaven abide unchanged."

In sum, then, when Copernicus' "postulates, which are called axioms", are correctly translated and properly understood, they provide a soundly reasoned and appropriately placed logical underpinning for the first presentation of the earth as a planet in motion.

NOTES

1. Noel M. Swerdlow: "The Derivation and First Draft of Copernicus's Planetary Theory; A Translation of the Commentariolus with Commentary," *Proceedings of the American Philosophical Society*, 1973, *117*, p. 437.

2. *Procli Diadochi in primum Euclidis Elementorum librum commentarii*, ed. G. Friedlein, Leipzig, 1873; reprint, Hildesheim, 1967; pp. 179: 2–5; 181: 8–10; 193: 15–17; 194: 4–7; 195: 17–19.
   Proclus de Lycie, Les Commentaires sur le premier livre des Eléments d'Euclide, tr. Paul Ver Eecke, Bruges, 1948, pp. 157, 159, 171, 172.
   Proclus, A Commentary on the First Book of Euclid's Elements, tr. Glenn R. Morrow, Princeton, 1970, pp. 140, 142, 152, 153.

3. Proclus, ed. Friedlein, p. 181: 17–21; Ver Eecke, p. 159; Morrow, p. 142.

4. Proclus, ed. Friedlein, p. 193: 20–194: 2; Ver Eecke, p. 171; Morrow, p. 152.

5. Benson Mates: *Stoic Logic*, 2nd printing, Berkeley, 1973, pp. 28, 132–133.

6. Aristotle: *Posterior Analytics*, I, 2: 72a17; I, 10: 76b14–15; in *Aristotelis opera*, 2nd ed., Berlin, 1960–1961.

7. G. Peano: *Opere scelte*, Rome, 1957–1959, III, pp. 116, 119.

8. D. Hilbert: *Foundations of Geometry*, tr. Leo Unger, 2nd ed., La Salle, 1971.

9. *The Medieval Science of Weights*, eds. E. A. Moody and M. Clagett, 2nd printing, Madison, 1960, p. 42.

10. Swerdlow, p. 438.

11. Swerdlow, p. 436.

12

When did Copernicus Write the Revolutions?

The 'Revolutions' (as *De revolutionibus orbium coelestium* may be called for the sake of brevity) inaugurated modern astronomy. For that reason the date of its composition by *Nicholas Copernicus* (1473–1543) is important for the history of science. But Copernicus was a taciturn man, who talked very little about himself and his work. For instance, he does not say precisely when he began to write the 'Revolutions'. Nevertheless, he does provide a clue.

He gives us this hint in an earlier and briefer treatise, now known as his 'Commentariolus'. "Here", he explains, "for the sake of brevity I have thought it desirable to omit the mathematical demonstrations intended for my larger work" [1]. The only "larger work" ever written by Copernicus is his 'Revolutions', in which he inserted the mathematical demonstrations omitted from his 'Commentariolus'. Evidently he was already planning the 'Revolutions', or perhaps he had already begun to write that "larger work", when he made this reference to it in his 'Commentariolus' [2].

As an alternative to this conclusion, a recent translator of the 'Commentariolus' said [3]:

I believe that the sort of book Copernicus was contemplating when he wrote the Commentariolus would have consisted of geometrical demonstrations of the equivalence of Ptolemy's and his own models.

Copernicus was absolutely convinced, however, that his own astronomy was right, and that Ptolemy's was wrong. Hence Copernicus never contemplated "geometrical demonstrations of the equivalence" of these two incompatible systems: the earth, for him, was a heavenly

---

[1] *Leopold Prowe:* Nicolaus Coppernicus. Bd. 2. Urkunden. Osnabrück 1967 (Neudruck der Ausgabe 1883–1884), S. 187: "Hic autem brevitatis causa mathematicas demonstrationes omittendas arbitratus sum maiori volumini destinatas."

[2] *H. Hugonnard-Roche, E. Rosen, J.-P. Verdet:* Introductions à l'astronomie de Copernic: Le Commentariolus de Copernic, la Narratio prima de Rheticus. Paris 1975, p. 74.

[3] *Noel M. Swerdlow:* The Derivation and First Draft of Copernicus' Planetary Theory, A Translation of the *Commentariolus* with Commentary. Proceedings of the American Philosophical Society 117 (1973), 439.

body in motion, whereas for Ptolemy it had been a non-heavenly body at rest.

The mistaken idea that Copernicus ever contemplated demonstrating the equivalence of the Ptolemaic and Copernican systems was promptly abandoned by this translator, Professor *Noel M. Swerdlow* of the University of Chicago. In place of his earlier error, Swerdlow soon substituted the obscurantist remark[4]:

What kind of larger book Copernicus had in mind when he wrote this [reference to the 'Revolutions' in his 'Commentariolus'] is by no means clear.

It is entirely clear, however, that the only "kind of larger book Copernicus had in mind" was the 'Revolutions'.

This identification of the 'Revolutions' as Copernicus' "larger book" is rejected by Swerdlow as a "hasty conclusion."[4] Far from being "hasty", however, this conclusion was the result of long and thoughtful reflection. As a carefully researched conclusion, it was first published nearly forty years ago[5]. Since that time it has been reprinted in two later and successively expanded editions of the same book (1959, 1971). In all that time it has evoked no disagreement on the part of any of the numerous reviewers of any of the three editions, nor on the part of any unprejudiced reader.

Yet, according to Swerdlow, "there is, in fact, no evidence whatsoever for" the conclusion that in the 'Commentariolus' Copernicus referred to the 'Revolutions'[4]. But there are in fact two excellent pieces of evidence for this conclusion. First, Copernicus refers to his larger work (maiori volumini); he planned and wrote only one larger work, namely, his 'Revolutions'. Secondly, he refers to the "mathematical demonstrations intended" (mathematicas demonstrationes . . . destinatas) for his larger work. Many mathematical demonstrations are included in his larger work, the 'Revolutions'. Nobody has ever raised any valid objection to the conclusion that in the 'Commentariolus' Copernicus refers to the 'Revolutions'. Certainly Copernicus never contemplated demonstrating the equivalence of his own system with Ptolemy's, nor did he ever have in mind any other kind of larger book than the 'Revolutions'.

Why, then, did Copernicus in the 'Commentariolus' refrain from referring to the 'Revolutions' by its title? Presumably he mentioned no title because he had not yet chosen any title. By the same token, when

---

[4] *N. M. Swerdlow:* The Holograph of the *Revolutions* and the Chronology of its Composition. Journal for the History of Astronomy 5 (1974), 188.

[5] *Edward Rosen:* Three Copernican Treatises. New York 1939, S. 59.

he circulated manuscript copies of his 'Commentariolus', it bore no title; the title *Commentariolus* was bestowed on it later by someone else. Copernicus decided to use 'Revolutions' as the title of his larger work only after he had finished writing the 'Commentariolus' and submitted copies of that untitled manuscript to a few trusted friends.

Just as Copernicus did not give the 'Commentariolus' any title nor sign his name as its author, so he failed to indicate when he wrote it. Fortunately, a copy of the 'Commentariolus' was listed in an inventory of the private library of a Cracow professor, who dated that catalogue of his books and manuscripts 1 May 1514[6]. "The *Commentariolus*", as Swerdlow correctly concludes, "must therefore have been written in early 1514 at the very latest."[7] Swerdlow also argues that "almost the entire contents of the *Commentariolus* seem to depend on [five] sources."[8] His fifth source is *Almagestum Cl. Ptolemei*, which was published on 10 January 1515. But how can a work published on 10 January 1515 have been a source for the *Commentariolus*, which "must ... have been written in early 1514 at the very latest"? By that time Copernicus was already planning the 'Revolutions' or perhaps had already begun to write it.

The Preface to the 'Revolutions' was written by Copernicus in June 1542[9]. In that Preface Copernicus says that a friend

asked me to publish this volume and finally permit it to appear after being buried among my papers and lying concealed not merely until the ninth year but by now the fourth period of nine years.[10]

A "fourth period of nine years" (quartum novennium) begins after 27 years. Subtracting 27 from 1542 leaves 1515 as the latest year in which "this volume" began to lie concealed among Copernicus' papers (apud me pressus). Copernicus plainly says that he concealed "this volume" (librum hunc), the 'Revolutions'. Yet Swerdlow misquotes Copernicus as saying that he "concealed his theories."[11] Swerdlow's twisting of *librum hunc* into "his theories" falsifies one of Copernicus' most famous utterances.

---

[6] *Ludwik Antoni Birkenmajer:* Stromata Copernicana. Cracow 1924, S. 199–202 (photofacsimile, S. 201).   [7] *N. M. Swerdlow:* (1973), 431.   [8] *Ibid.* S. 425.

[9] *Enrico Stevenson, Jr.:* Inventario dei libri stampati palatino-vaticani. Vol. 1, part 2. Rome 1886–1889, S. 161, N. 2250.

[10] *Nicolai Copernici* Opera omnia. Vol. II. Warsaw/Cracow 1975, S. 3: ". . . efflagitavit, ut librum hunc ederem et in lucem tandem prodire sinerem, qui apud me pressus non in nonum annum solum, sed iam in quartum novennium, latitasset."

[11] *N. M. Swerdlow* (1974), 191.

Swerdlow not only misquotes and falsifies what Copernicus said, but he also misrepresents Copernicus' professional behavior. For, according to Swerdlow, Copernicus "concealed his theories" throughout "the entire period that had elapsed since his first discovery of the heliocentric theory." [11] Now Swerdlow contends "that the *Commentariolus* was written in haste, possibly very shortly after Copernicus developed his new theories." [12] But if Copernicus "concealed his theories" throughout "the entire period that had elapsed since his first discovery of the heliocentric theory", how did a copy of Copernicus' 'Commentariolus' reach a Cracow professor before 1 May 1514?

According to the earliest well-informed biographer of Copernicus [13],

he had as friends . . . Cracow astronomers, formerly his fellow-students, with whom he corresponded about eclipses and observations of eclipses, as is clear from the letters written by his own hand which are in the possession of Jan Brozek in the University of Cracow.

With one exception, the letters written by Copernicus to his astronomical friends in Cracow have unfortunately not survived. Nevertheless, those lost letters suggest the answer to the question how a copy of Copernicus' 'Commentariolus' reached a Cracow professor before 1 May 1514:

Copernicus must have sent one or more copies of his 'Commentariolus' to Cracow. At that time, far from concealing his heliocentric theory, he disclosed it to his professional friends. His conduct disproves Swerdlow's misstatement that he "concealed his theories" throughout "the entire period that had elapsed since his first discovery of the heliocentric theory."

While disclosing his heliocentric and geokinetic theory in the 'Commentariolus', Copernicus took care not to publicize himself as the author. Hence his little treatise was inventoried on 1 May 1514 as an anonymous manuscript. Why did Copernicus circulate his 'Commentariolus' anonymously before 1 May 1514, and then begin to conceal his 'Revolutions' not later than 1515? Did some hostile reaction to the 'Commentariolus' intensify the fear he had already felt when he

---

[12] *N. M. Swerdlow* (1973), 429.

[13] *Szymon Starowolski:* Hekatontas. Venice 1627, S. 158–162. Quoted by *Franz Hipler:* Nikolaus Kopernikus und Martin Luther. In: Zeitschrift für die Geschichte und Alterthumskunde Ermlands 4 (1867–1869), 538. Also quoted by *L. A. Birkenmajer:* Mikołaj Kopernik. Cracow 1900, S. 453: "familiares habuit . . . mathematicos Cracovienses, olim condiscipulos suos, cum quibus conferebat de eclipsibus et earum observationibus, ut patet ex epistolis manu illius ipsius scriptis, quas habet in Academia Cracoviensi Jo. Broscius."

decided to hide his authorship of that earliest extended astronomical treatment of the earth as a moving planet?

We have now examined the only two statements ever made by Copernicus about the time when he began to write the 'Revolutions'. Before 1 May 1514, in the 'Commentariolus', he refers to the 'Revolutions' as already planned or begun. In June 1542, in the Preface to the 'Revolutions', he states that he has kept the 'Revolutions' hidden since 1515. Copernicus wrote these two statements some thirty years apart. Nevertheless they are in substantial agreement. They are both expressed in simple and straightforward Latin. When they are read in a simple and straightforward manner, undistorted by any preconception, their agreement tells us that Copernicus began to write the 'Revolutions' not later than 1515.

The preconception responsible for Swerdlow's distortion of these two simple passages is his contention that

in all probability the *Revolutions* was entirely a work of the 1530's, and I can see no evidence for giving any part of it an earlier date [14].

Let us now examine some evidence that Swerdlow says he cannot see. In June 1542, in the Preface to the 'Revolutions' Copernicus recalled:

Not so long ago under Leo X the Lateran Council considered the problem of reforming the ecclesiastical calendar. The issue remained undecided then only because the lengths of the year and month and the motions of the sun and moon were regarded as not yet adequately measured. From that time on, at the suggestion of that most distinguished man, Paul, bishop of Fossombrone, who was then in charge of this matter, I have directed my attention to a more precise study of these topics.

Paul of Middelburg (1445–1533), bishop of Fossombrone, published a report to Leo X about the outcome of that pope's efforts to stimulate projected corrections of the defects in the current calendar. Among the astronomers who wrote in response to the pope's invitation, Paul of Middelburg listed Copernicus [15]. But what Copernicus wrote on that occasion has not been found. It is unmistakably connected, however, with the *Revolutions*, III, 16.

There Copernicus reminds his reader that Ptolemy "thought that for the rest of time" the earth-sun eccentricity and the direction of the earth-sun apsidal line "would abide forever". But Copernicus adds at once: "Both values are now found to have changed with a perceptible difference". Yet when Copernicus wrote the 'Commentariolus', he still

---

[14] *N. M. Swerdlow* (1974), 194.

[15] *Paul of Middelburg:* Secundum compendium correctionis calendarii. Rome 1516, Sig. blr.

believed this eccentricity and this direction to be invariable. When did he discover their variability? In the *Revolutions,* III, 16, he says:

In the more than ten years since I have devoted my attention to investigating these topics, and particularly in 1515 A.D., I have found . . .

Now Copernicus replied to *Paul of Middelburg* in or about 1515. What he found "particularly in 1515 A. D." (praesertim anno Christi MDXV.) made him modify two of the main tenets upheld in his 'Commentariolus': the invariability of the earth-sun (a) eccentricity and (b) apsidal direction. Copernicus' dating of the composition of *Revolutions,* III, 16, is not as precise as we might wish. But when he there speaks of "the more than ten years since I have devoted my attention to investigating these topics, and particularly in 1515 A. D." (a decem et pluribus annis, quibus earum rerum perscrutandarum adiecimus animum, ac praesertim anno Christi MDXV.), he surely implies that he wrote *Revolutions,* III, 16, before 1530.

Let us now look at some more evidence that Swerdlow says he cannot see. In *Revolutions,* I, 10, Copernicus recalls "that, according to Ptolemy, there are 38 earth-radii to the moon's perigee". "But", Copernicus adds at once, "according to a more accurate determination there are more than 49, as will be made clear below". Below, in the *Revolutions,* IV, 17, 22, 24, Copernicus put the moon's perigeal distance at more than 52 earth-radii. After he had obtained this later result (>52), in his characteristic manner he forgot to go back to I, 10, in order to raise "more than 49" there by 3. The resulting inconsistency between I, 10, and IV, 17, 22, 24, was removed in the first edition of the 'Revolutions' (Nuremberg, 1543). But it remains in Copernicus' autograph [16].

It is of course true that 52 is more than 49, and perhaps someone may think that ">52" in IV, 17, 22, 24, carries out the promise in I, 10, to make "clear below" that the moon's perigeal distance is ">49, according to a more accurate determination". However, had Copernicus already possessed the ">52" result when he was writing I, 10, surely he would then have said ">52" rather than ">49". For, the larger number would have strengthened his objection in I, 10, to the reasoning of the Ptolemaists. His use of ">49" in I, 10, proves that when he wrote I, 10, he did not yet have the ">52" result.

That bigger value was derived from the calculations based on the two observations discussed in the *Revolutions,* IV, 16. These two ob-

---

[16] *Nicholas Copernicus:* Complete Works. Vol. I. The Manuscript of Nicholas Copernicus' On the Revolutions. Facsimile. London/Warsaw 1972, Fol. 8v, line 14.

servations are dated 27 September 1522 and 7 August 1524. Therefore Copernicus wrote I, 10 (and folio 8 in his autograph) before he drew his conclusions from these observations of 1522 and 1524. Folio 8 was written on paper C, one of the four batches of paper used by Copernicus for his autograph. The observations of 1522 and 1524 are reported on folios 126 and 127, which belong with paper D, another of Copernicus' four batches. He generally used paper C before paper D, as he did here, but the opposite order also occurs. In any case, he wrote folio 8 and *Revolutions,* I, 10, long before 1530.

On folio 11, in *Revolutions,* I, 11, Copernicus remarked that, in addition to the eight celestial spheres accepted since ancient times,

some people . . . adopted a ninth surmounting sphere. This having proved inadequate, more recent writers now add on a tenth sphere.

When Copernicus later returned to this subject in *Revolutions,* III, 1, he had still more recent information:

Some people excogitated a ninth sphere, and others a tenth, by which they thought that these phenomena [nonuniform precession of the equinoxes, diminution in the obliquity of the ecliptic] are brought to pass in this way. Yet they could not furnish what they promised. An eleventh sphere too has already begun to emerge into the light of day.

When did Copernicus, in III, 1, learn of an eleventh sphere, about which he knew nothing in I, 11?

A Cracow correspondent of Copernicus sent him a copy of Johann Werner's *Motion of the Eighth Sphere* (Nuremberg, 1522) and asked his opinion of that work. While perusing it, Copernicus found out that "an eleventh sphere too has already begun to emerge into the light of day". Hence he wrote the passage in I, 11, before he saw Werner's *Motion of the Eighth Sphere.* For, had Werner's eleventh sphere been known to Copernicus when he was writing I, 11, he would surely have fortified his position there by mentioning an eleventh sphere. Instead, in I, 11, he says: "more recent writers now add on a tenth sphere".

Clearly, Copernicus wrote I, 11, before he read Werner, and he wrote III, 1, after he read Werner. He dated his *Letter against Werner* 3 June 1524[17]. Hence he wrote *Revolutions,* I, 11, before that date, which is of course much earlier than 1530.

In opposition to this conclusion, Swerdlow argues that

Werner uses only an eighth, ninth, and tenth sphere in his theory of irregular precession and change of obliquity. While there is an additional sphere for the diurnal

---

[17] *E. Rosen:* Three Copernican Treatises, 3rd ed., New York 1971, S. 106.

rotation of the entire universe, it is always called the *primum mobile,* and would not be counted as an eleventh sphere in considering the theory of precession [18].

Neither would the first seven spheres – those of the planets – be counted in considering the theory of precession. In counting the total number of spheres, however, like other sixteenth-century astronomers Copernicus considered not only the theory of precession but the entire structure of the cosmos. No sphere was omitted from this count, least of all the sphere which caused the daily rotation, that most prominent of all celestial phenomena – the apparent daily rising and setting of the sun, moon, and stars.

Since Swerdlow refuses to count the sphere responsible for the daily rotation, for him "it is difficult to know what the eleventh sphere in III, 1, means" [18]. In like manner, "What kind of larger book Copernicus had in mind when he wrote this [reference to the 'Revolutions' in his 'Commentariolus'] is by no means clear" to Swerdlow, as we saw above.

Although Swerdlow finds it "difficult to know what the eleventh sphere in III, 1, means", he suggests that Copernicus' reference there to an eleventh sphere

is a slightly sarcastic remark about the great number of spheres employed to account for these motions [18].

But there is nothing sarcastic in Copernicus' next remark in III, 1:

By invoking the motion of the earth, I shall easily refute this number of circles as superfluous by showing that they have no connection with the sphere of the fixed stars.

Swerdlow also argues that "the passage in I, 11, must itself be a reference to Werner's theory". [18] Yet in I, 11, Copernicus makes no reference to the theory of any single individual. Instead, he mentions "more recent" (recentiores) writers in the plural, who "now add on a tenth sphere" (nunc . . . decimam superaddunt). These unspecified adherents to a ten-sphere cosmos preceded the introduction of the eleventh sphere mentioned in III, 1. The ten-sphere cosmos was familiar, says Copernicus in III, 1, before an eleventh sphere "began (coeperat) to emerge into the light of day". Copernicus' use of *Iam . . . coeperat* in III, 1, shows that he is there referring to a development that began shortly before he wrote III, 1. The eleventh sphere was introduced (as far as Copernicus knew) by *Werner* in 1522. Copernicus

[18] *N. M. Swerdlow:* On Copernicus' Theory of Precession. In: *Robert S. Westman,* ed.: The Copernican Achievement. Berkeley/Los Angeles/London 1975, S. 97.

first learned about Werner's eleventh sphere before 3 June 1524. Hence he wrote III, 1, about that time, in any case long before 1530.

A younger contemporary of Copernicus was less reticent than he was in naming adherents of the ten-sphere cosmos and the initiator of the eleven-sphere cosmos. In his *Commentaries on George Peurbach's New Theories of the Planets* (Basel, 1566), *Erasmus Oswald Schreckenfuchs* (1511–1579), professor of astronomy at the University of Freiburg-im-Breisgau, referred to those who "conceived two spheres above the eighth [sphere], as was done by *Peurbach*, the author of the 'Theories', in imitation of the opinion of the Alfonsine [Tables]". Then, in discussing "Johann Werner, a citizen of Nuremberg", Schreckenfuchs stated:

In order to account in a convenient way for the irregularity of the equinoxes and of the sun's maximum inclination, Werner added an extra sphere on top of the spheres of the Alfonsine [Tables] [19].

The authors of the canons prefixed to the Alfonsine Tables were supporters of the ten-sphere cosmos, as was Peurbach, according to Schreckenfuchs, who attributed the introduction of an eleventh sphere to Werner. The absence of Werner's eleventh sphere from *Revolutions*, I, 11, proves that there is no "reference to Werner's theory" there, despite Swerdlow. Copernicus wrote I, 11, before 1524, at the very least half a dozen years before 1530.

As his final argument in this context Swerdlow points out [18] that I, 11, on folio 11 in Copernicus' autograph "is a sheet of paper of watermark D"; on the other hand, III, 1, on folio 71 "is of watermark C". "Therefore", Swerdlow concludes, "the passage in I, 11, must be later than or perhaps contemporary with III, 1, but cannot be earlier". Yet Swerdlow recognizes that "the order of use [by Copernicus of his four batches of paper] was not as simple as first C, then D, then E, for there is obviously some overlapping with two papers being used simultaneously". [20] In other overlapping, D was used before C, as in this instance of folio 11 (paper D) being used before folio 71 (paper C). Evidently Copernicus did not completely exhaust his stock of paper C before

---

[19] *E. O. Schreckenfuchs:* Commentaria in novas theoricas planetarum Georgii Purbachii. Basel 1566, S. 388–389: ". . . imaginati sunt duas spheras supra octavam, sicut fecit Purbachius autor Theoricarum, qui Alphonsinorum imitatus est sententiam . . . Nam Vernerus, ut commode conservaret diversitatem et aequinoctiorum et maximae solis declinationis, addidit adhuc unam sphaeram ultra sphaeras Alphonsinorum."   [20] *N. M. Swerdlow* (1974), 189.

using any of paper D. That is why D sometimes precedes C in Copernicus' autograph.

As we saw above, *Revolutions,* I, 11, on paper D, was written before 3 June 1524; III, 1, on paper C, about 1524; I, 10, before 1524; and III, 16, shortly after 1525. Yet, according to Swerdlow, "in all probability the 'Revolutions' was entirely a work of the 1530s, and I can see no evidence for giving any part of it an earlier date".[14] The foregoing evidence (that Swerdlow says he cannot see) harmonizes well with Copernicus' indication, in or before 1514, that he is planning or has begun the 'Revolutions', and with his recollection in 1542 that the 'Revolutions' had been hidden since 1515.

Swerdlow's contention that no part of the 'Revolutions' is earlier than 1530 wipes out at least sixteen years of the three periods of nine years during which Copernicus says he kept the 'Revolutions' hidden. Those sixteen lost years include Copernicus' forties and early fifties, normally a productive period for a healthy intellectual like Copernicus. He performed most of his dated observations during those sixteen allegedly lost years. Those were not random observations. On the contrary, they were all undertaken to confirm or disconfirm some particular part of his geokinetic theory. This underwent a radical transformation after 1514, so that the 'Commentariolus' is really only the first stage in the development of Copernicus' geokineticism, while the 'Revolutions' is the second and mature stage. The turning point between the two stages may have been what Copernicus found "particularly in 1515". In the following ten or more years he was busy writing *Revolutions,* Book III. On 3 June 1524 he ended his 'Letter against Werner' with the question[17]:

What finally is my own opinion concerning the motion of the sphere of the fixed stars? Since I intend to set forth my views elsewhere, I have thought it unnecessary and improper to extend this communication further.

Copernicus' views concerning the motion of the sphere of the fixed stars are set forth in *Revolutions,* III, 1–12. Those views were written or planned by 3 June 1524.

As we consider the foregoing mass of evidence, we have every reason in good faith to accept what Copernicus says and implies about the time when he began to write the 'Revolutions': about 1515. He worked at that long and difficult volume steadily for nearly three decades. When did he finish writing the 'Revolutions'?

His disciple *George Joachim Rheticus* (1514–1574), while visiting him, "wrote from Prussia that he was waiting for his teacher [Copernicus] to finish his work" [the 'Revolutions'][21]. This message was transmitted to *Philip Melanchthon* on 15 April 1541[22]. Seven weeks later, on 2 June, while still waiting, Rheticus reported that Copernicus "is enjoying quite good health and is writing a great deal".[23] Then, a week later, on 9 June, as a dinner guest Copernicus confided to his host that "he had finally overcome his prolonged reluctance to release his volume for publication".[24] On 29 August, when the fair copy was ready for the printer, Rheticus asked the duke of Prussia to send letters to the Elector of Saxony and the University of Wittenberg, where he was a professor, requesting that he "be permitted to publish his teacher's work".[25] Three days later, on 1 September, the duke's secretary wrote similar letters to the Elector and to the University on behalf of Rheticus: "He is thinking of publishing outside of Prussia a book in his field, which he put together and finished in this region with great energy, effort, and labor".[26] What was "put together and finished . . . with great energy, effort, and labor" on Rheticus' part was, presumably, the fair copy of Copernicus' 'Revolutions'. In the twin letters of 1 September the ducal secretary also asked the Elector and the University to grant Rheticus time to publish the book on paid leave, "without interruption of his salary as a professor".[27] His status immensely strengthened by these unconditional endorsements by the duke of Prussia, Rheticus returned to the University of Wittenberg, where he was elected dean of the liberal arts faculty on 18 October 1541. At the end of the winter semester on 1 May 1542, he left Wittenberg for Nuremberg in order to supervise the printing of Copernicus' 'Revolutions'

[21] *Karl Heinz Burmeister* in: Isis 61 (1970), 386: "Magister Ioachimus scripsit ex Prussia, se expectantem absolutionem operis sui praeceptoris . . ."

[22] Corpus reformatorum. *Philippi Melanthonis* Opera. Halle/Braunschweig 1834–1860. IV, col. 174.

[23] *K. H. Burmeister:* G. J. Rhetikus. Wiesbaden 1967–1968. Bd. III, S. 27: "satis commode valet et multa scribit."

[24] *E. Rosen:* Three Copernican Treatises, 3rd ed. S. 371.

[25] *K. H. Burmeister* (1967–1968) III, 39: "mir vergonnet mochte werden, das opus D. praeceptoris mei in den truk zu geben."

[26] *Ibid.* III, 40: "ehr ein buch seiner kunst, welches ehr alhie In diesen landen mit grossem vleiss, muhe und arbeitt zusamen gepracht und verfertiget offentlich In druck draussen landes ausgehen zulassen bedacht."

[27] *Ibid.:* "Ime genediglichen gestatten und vergonnen, das ehr sich zw volfurung solches seines vorhabenden wercks an die orth da ehr sein buch trucken zulassen entschlossen ein zeitlang ohne abbruch seiner besoldung der lectur begeben moege."

there. Clearly, then, the writing of the 'Revolutions' was finished by Copernicus in the previous summer, perhaps by 9 June 1541.

Swerdlow combats "the old notion that Copernicus had completed his work in 1530 or so".[20] Perhaps Swerdlow is now resurrecting *Jean Czynski* (1801–1867), who in 1847 mistakenly said that Copernicus "finished his great work on the *Revolutions* in 1530".[28] But this is hardly the time, after thirteen decades, to do battle with a hapless exile from the ill-fated Polish insurrection of 1830. In the main Czyński published passionate patriotic, religious, and humanitarian tracts, while advocating the liberation of the serfs and the emancipation of the Jews. Toward Copernicus, his attitude was less scholarly than hagiographical. Now is not the time to emulate his pitiful blunders concerning Copernicus.

13

Copernicus' Spheres and Epicycles

Nicholas Copernicus (1473–1543), the founder of modern astronomy, still used the ancient device of the interacting deferent and epicycle to account for the planetary motions. He continued to employ such combinations of circles and spheres, without anticipating that his great follower Johannes Kepler was going to discover the ellipticity of the planetary orbits and thereby banish the sphere and the epicycle forever from this aspect of science.

For the three outer planets (Mars, Jupiter, Saturn) known in his time, in his *Commentariolus* Copernicus indicated the orbit's relative size by assigning a number to the deferent-radius (*semidiameter orbis*). This term was defined by Copernicus, immediately after he had given those three numbers, in the following way: *Dico autem semidiametrum* [*orbis* is to be supplied here] *a centro orbis ad centrum primi epicycli distantiam* (By '[deferent-]radius' I mean the distance from the deferent's center to the center of the first epicycle); 'first epicycle,' Copernicus at once proceeded to explain, because there was a second epicycle for each of these planets in the *Commentariolus*.

Copernicus' definition of deferent-radius was recently translated as follows: 'I call the *semidiameter* the distance from the center of the sphere to the center of the first epicycle' ([6], 465). This translator, Dr. Noel M. Swerdlow, Assistant Professor of History at the University of Chicago, then remarks: 'If he [Copernicus] were talking only about circles and epicycles, this distinction would be unnecessary since the center of the epicycle is obviously located at the circumference of the circle on which it moves' ([6], 466). But no distinction was made here by Copernicus. Instead, he gave a simple definition, which he must have felt was needed by his readers in the absence of a diagram.

What is 'this distinction' imagined by Swerdlow? His remarks continue as follows:

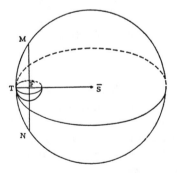

Swerdlow's Fig. 12.

With solid spheres, however, the radius of the solid sphere itself must extend to the outer edge of the epicycle, or better, to the outer surface of the epicyclic sphere. The arrangement is shown in figure 12. The radius of the solid sphere is $\bar{S}T$, and the epicyclic sphere with center **F** is tangent to the larger sphere at T and presumably rotates on an axis attached to the solid sphere at M and N. Hence Copernicus must specify that he is calling $\bar{S}F$, not $\bar{S}T$, the semidiameter of the sphere. ([6], 466–467)

How can the sphere's semidiameter be $\bar{S}F$, which falls short of the sphere's surface? The attribution of such mathematical stupidity to the founder of modern astronomy is possible only for someone like Swerdlow, who holds a very low opinion of Copernicus' mental ability.

In Swerdlow's Figure 12, the sphere centered at $\bar{S}$ is a 'solid sphere.' Despite its solidity this sphere, according to Swerdlow, wholly contains the first epicyclic sphere. Moreover, 'the second epicyclic sphere is entirely contained within the first in the same way' ([6], 467). The second epicyclic sphere in turn carries the moving planet, itself a solid sphere. Whatever may be thought of this highly constipated conception, it does not in the least resemble the planetary theory expounded in Copernicus' *Commentariolus*.

That theory is explained by Swerdlow with the aid of his Figure 14. Certain features irrelevant to the present discussion have been removed from his Figure 14, which is reproduced here in a simplified form.

Swerdlow's Fig. 14 (somewhat simplified).

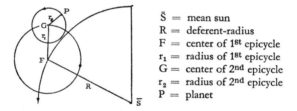

$\bar{S}$ = mean sun
R = deferent-radius
F = center of 1st epicycle
r_1 = radius of 1st epicycle
G = center of 2nd epicycle
r_2 = radius of 2nd epicycle
P = planet

Swerdlow says ([6], 468): 'we let $\bar{S}$ be . . . the mean sun in Copernicus's model.' The term 'mean sun' (as contrasted with 'true sun') usually refers to the sun's uniform motion. But Copernicus' sun had no motion, mean or true ('What appear to us as the sun's motions are not due to its motion but to the earth's motion,' *Commentariolus*, Axiom 6). Swerdlow, however, has his own definition of 'mean sun,' which is his label for what Copernicus calls the center of the earth's annual orbit. ([6], 442) Both the term 'mean sun' and the corresponding concept are thoroughly alien to Copernicus' thinking.

Swerdlow's Figure 14, thus simplified, shows at a glance that the planetary theory actually propounded in Copernicus' *Commentariolus* requires the intersection of the two epicycles with each other and of both epicycles with the deferent. But, according to Swerdlow ([6], 467), 'The intersection of spheres is not permitted' by Copernicus. Since intersection, as shown by Swerdlow's own Figure 14, was an indispensable part of Copernicus' planetary theory in the *Commentariolus*, that theory did not use solid spheres.

What then did it use? The *Commentariolus*' first *Axiom* is *Omnium orbium caelestium sive sphaerarum unum centrum non esse*. On his translation: 'There is no one center of all the celestial spheres (*orbium*) or spheres (*sphaerarum*),' Swerdlow comments: 'I take *sphaerarum* to be a synonym clarifying the meaning of *orbium*' ([6], 436, 438). *Sphaerarum* and *orbium* are the genitive plural forms of *sphaera* and *orbis*, two key terms in Copernicus' technical vocabulary. *Sphaera* is not exactly a synonym of *orbis*, nor is *sphaera* used here to clarify the meaning of *orbis*. Copernicus took for granted that his readers would be thoroughly familiar with the distinction between *orbis* and *sphaera*. But since that distinction is evidently no longer clear to everybody nowadays, let us recall how it was explained in a textbook of elementary mathematics by a slightly younger contemporary of Copernicus:

Among solid figures, the most important is the sphere (*sphaera*), [which is] the most regular of all. It is a regular solid body, bounded by a single surface ... We conceive the sphere to be generated by the complete rotation of a semicircle: while the diameter of the semicircle remains fixed, the plane surface of that circle is rotated ... An orb (*orbis*) is also a solid figure. It is bounded, however, by two round spherical surfaces, namely, an interior [surface], which is called concave, and an exterior [surface], which is labeled convex. If these surfaces have the same center, the orb will be uniform, that is, of equal thickness throughout. But if the surfaces have different centers, these will make the orb's thickness nonuniform and irregular. This is the kind which the heavens of all the planets have.

Inter solidas figuras primatum habet sphaera omnium regularissima, quae est corpus solidum regulare, unica superficie terminatum ... Imaginamur autem describi sphaeram ex completo semicirculi circumductu, cum uidelicet semicirculi diametro manente fixa, eiusdem circuli plana superficies circumducitur ... Orbis est quoque figura solida, duabus tamen rotundis sphaericisque superficiebus terminata, utpote interiori, quae concaua dicitur, & extrinseca quae conuexa nominatur. Harum superficierum si idem fuerit centrum, orbis ille erit uniformis, id est, aequalis undique crassitudinis. Sin diuersa centra ipsae superficies habuerint, efficient difformem & irregularis crassitudinis orbem, cuiusmodi habnt omnium planetarum coeli ... ([3], 60)

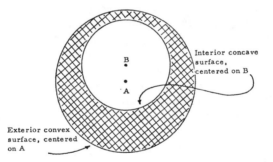

Münster's Diagram for an Orb of Nonuniform Thickness. Cross-hatching indicates the orb's solid body. The space enclosed within the orb's solid body is hollow.

This distinction between *sphaera* and *orbis*, between solid sphere and nonsolid orb, discloses the fundamental error in Swerdlow's unsupported assertion: 'Copernicus had no doubt that the motions of the planets were controlled by solid spheres' ([6], 466).

In the *Commentariolus*, after giving the three outer planets two epicycles each, Copernicus recalled that in his lunar theory the moon likewise had two epicycles, with the moon attached to the second epicycle. This second lunar epicycle was used by Copernicus to diminish the grossly excessive variation of the moon's apparent diameter in Ptolemy's lunar theory. Hence the eminent Portuguese mathematician and astronomer Pedro Nunes (1492–1577), otherwise a confirmed anti-Copernican, said:

> Not without reason does [Copernicus] place the moon on an epicycle of an epicycle, with the center of the smaller [epicycle] on the circumference of the larger [epicycle]. I point out, however, that the entire smaller [epicycle] must be enclosed within the larger [epicycle] to avoid shattering the heaven.
>
> *Lunam non sine ratione collocat in epicyclo epicycli, centrum minoris in circumferentia maioris. Caeterum aduerto totum minorem intra maiorem includi oportere, ne coelum rumpatur* ... ([4], 106)

Not only was Nunes worried that his heaven would be shattered by Copernicus' protruding second lunar epicycle, but the Portuguese astronomer was also uneasy about the heavenly spaces left vacant by Copernicus. What Nunes read was not Copernicus' *Commentariolus*, but his mature *Revolutions* which, by contrast with the earlier *Commentariolus*, made the planetary deferent orbs eccentric, that is, their centers were at a certain distance from the center of the universe. Consequently, Nunes remarked:

> Since [Copernicus] adopts eccentric orbs, he will therefore have to assume others, in order to fill out the planetary spheres concentric with the universe.
>
> *Et quoniam eccentricos orbes ponit: alios igitur ponere necesse erit, qui planetarum sphaeras mundo concentricas compleant.* [(4], 106)

Unaware that the author of the *Revolutions* was already dead, Nunes warns Copernicus that he will have to resort to solid spheres. Nunes thereby clearly implies that Copernicus did not do so. Himself an outstanding sixteenth-century scientist, Nunes understood Copernicus' thinking. Swerdlow does not.

Consider, for example, Swerdlow's misunderstanding of a simple yet crucial word in the *Commentariolus*, where Copernicus expressed confidence that he would provide powerful proof of the earth's motion 'in his exposition of the circles' of the heavenly bodies. His expression here (*in circulorum declaratione*) echoed his announcement in the preceding sentence that he would give the 'lengths of the radii of the spheres in the explanation of the circles themselves' (*Quantitates tamen semidiametrorum orbium in circulorum ipsorum explanatione hic ponentur*). Swerdlow translates: 'Nevertheless, the lengths of the semidiameters of the spheres will be set down here in the explanation of their circles' [(6], 438]. These last two words, 'their circles,' undoubtedly mean 'the spheres' circles.' Had this been what Copernicus meant, he would certainly not have written *circulorum ipsorum*, which is not how 'their circles' was expressed in Latin.

By his deliberate choice of the words *circulorum ipsorum* Copernicus emphasized that in his astronomical thinking the 'circles themselves' were foremost. These circles were the most salient features of Copernicus' nonsolid spheres. That is why Copernicus announced, as we just saw, that he would give the 'lengths of the radii of the spheres in the explanation of the circles themselves.' Despite Swerdlow, Copernicus' cosmos did not consist of solid spheres. These could not intersect. But Copernicus' circles could and did.

Swerdlow's misconception that Copernicus' cosmos consisted of solid spheres leads him to misinterpret an important statement in the *Commentariolus*. There, just before giving the relative sizes of the orbits of the three outer planets, Copernicus suggested that those sizes were linked with the periods of revolution of these planets: Saturn had the largest orbit and the longest period; Mars had the smallest orbit and the shortest period; and Jupiter's orbit and period came between the other two. Copernicus' term for size (*magnitudo*) referred only to a dimension, as was indicated very clearly by his selection of the connecting word 'For' (*Nam*), which linked 'size' with the three numbers discussed at the beginning of this article. Although Copernicus restricted his statement to immaterial dimensions, Swerdlow fancies that Copernicus 'fell back on the possibility that the sheer massiveness of the spheres retarded the planetary motions' ([6], 466). Swerdlow also speaks of 'the very mention of the possibility that the spheres have some kind of mass' ([6], 466). This possibility is mentioned by Swerdlow. It was never mentioned by Copernicus.

Here Swerdlow attributes to Copernicus the idea that mass is a property of the heavenly spheres (as distinguished from the physical bodies of the planets and stars). Elsewhere, however, Swerdlow implies that Copernicus could not even have entertained any such idea. For, after mentioning 'the motion proper to the substance of the heavens,' Swerdlow decides that 'Speculations about such things do not belong to the domain of mathematical astronomy,' since such speculations belong to the domain of natural philosophy ([6], 435). Then, ignoring Copernicus' undeniably solid contributions to the basic concepts of natural philosophy or physics, Swerdlow classifies him as 'a pure astronomer,' 'an astronomer, not a natural philosopher,' and adds: 'To understand Copernicus' work properly, as he understood it, one must completely remove it from natural philosophy' ([6], 440). If so, one must completely remove from Copernicus' work, properly understood in Swerdlow's manner, the conception of massive heavenly spheres.

These spheres, whether massive or not, induce Swerdlow to be false to his own translation of Copernicus. For, Swerdlow argues that

> it was precisely this . . . concern for the representation of the motions of the planets by the rotation of spheres that led Copernicus . . . to investigate alternatives to Ptolemy's planetary models. At the beginning of the *Commentariolus* Copernicus explains that this was indeed the original motivation of his researches in planetary theory. In the course of laboring over this 'exceedingly difficult and nearly insoluble problem' he found something vastly more important than the strict representation of planetary motions by rotating spheres. . . . ([6], 467)

Yet this passage at the beginning of the *Commentariolus* is translated by Swerdlow himself as follows:

> I often pondered whether perhaps a more reasonable model composed of circles could be found from which every apparent irregularity would follow while everything in itself moved uniformly, just as the principle of perfect motion requires. ([6], 435)

Copernicus' key term here is *circulorum*, translated as 'circles' by Swerdlow. But when Swerdlow later harks back to the beginning of the *Commentariolus*, he distorts his own translation by substituting spheres for Copernicus' circles.

To describe the aim of his rendering, Swerdlow says: 'The translation is done strictly' ([6], 432). As emendations in the Latin text of the *Commentariolus*, this strict translator seeks to couple the preposition *cum* with the accusative case instead of the ablative, and to make the accusative *diminutionem* the subject of a proposed verb *asseritur* ([6], 455, 461). Exhibiting such impeccable mastery of Latin, this strict translator has decided that certain propositions which Copernicus 'called axioms' were 'incorrectly called axioms' ([6], 436, 437). This strict translator condemns as 'meaningless' the word *asse*, used by Copernicus ([6], 505). But having received his doctoral degree in law, Copernicus was surely familiar with the distinction *aut ex asse aut pro parte* ('either entirely or partly') ([1], II, 8, 15, 1). Swerdlow's translation describes the moon's smaller epicycle as 'observant' ([6], 490). Yet nowhere in nearly a hundred pages devoted to discussing such matters does Swerdlow deign to explain what he means by an 'observant epicycle' and how it would differ from an inobservant epicycle.

Ptolemy's equant was rejected by Copernicus. In his translation of the objection to the equant, Swerdlow makes Copernicus say: 'the planet never moves with uniform velocity' ([6], 434). But Copernicus never said any such thing. For if a planet moves with nonuniform velocity, in a pattern repeated over and over again, its observed nonuniform motion is sometimes faster, and sometimes slower, than its computed uniform motion. However, as it passes from faster-than-uniform to slower-than-uniform, as also in the opposite direction, the planet must move with its uniform velocity. 'The planet never moves with uniform velocity' is a gross error found only in Swerdlow. By contrast, Copernicus rejected Ptolemy's equant because it made 'the planet move with a velocity that was not always uniform' (*neque . . . aequali semper velocitate sidus moveri*).

In order to account for the earth's annual revolution around the sun, Copernicus introduced what he called the *orbis magnus*. This term, which was later adopted by the Copernicans as a war cry in their indomitable struggle against the vested and institutionalized geocentric and geostatic astronomy, is translated by Swerdlow as 'great sphere' ([6], 441). More generally, Swerdlow asserts: 'In the *Commentariolus orbis* always means sphere' ([6], 432). In that work Copernicus explained that a certain phenomenon reaches its maximum when an outer 'planet is seen along the line of sight tangent to' *circumferentiam magni orbis*; 'the circumference of the great sphere,' Swerdlow's translation continues ([6], 480). But a straight line is tangent to the circumference of a circle, not a sphere. In explaining his own Figure 26, Swerdlow himself says: 'we describe a circle of radius *s* carrying the earth' ([6], 482). Swerdlow's 'circle . . . carrying the earth' is Copernicus' *orbis magnus*.

> The reason for using this expression [*orbis magnus*, says Swerdlow] is never given [by Copernicus]. Kepler, who usually understands Copernicus's motivation better than Copernicus himself, clearly does not know why when he guesses, both in the *Mysterium Cosmographicum* and in the *Epitome Astronomiae Copernicanae*, that the earth's orbit is called the Great Sphere because it has so many uses. ([6], 442)

Does Swerdlow understand Kepler any better than he understands Copernicus? In his *Mysterium Cosmographicum*, on a page cited by Swerdlow, Kepler gives a diagram of the universe in which the smallest circle is Mercury's (*minimus . . . circulus est Mercurii*). After Venus' circle, there follows 'AB, the earth's . . . It is called ORBIS MAGNUS, on account of its multiple uses' ([2], I, 20 = VIII, 36).

On the other page of the *Mysterium Cosmographicum* cited by Swerdlow, Kepler exclaims: 'So numerous and so important are these results, which were achieved by Copernicus through the location and motion of circle AB alone, that he rightly called it MAGNUS, even though it was small' ([2], I, 19 = VIII, 38). In his *Epitome Astronomiae Copernicanae* Kepler gave the following question and answer:

> In the astronomy of Copernicus what is the *orbis magnus*? This is the name applied by Copernicus to the true orbit (*orbita*) of the earth about the sun. This orbit is located in the space between the orbit of Mars outside it and the orbit of Venus within it; and he calls it *magnus* not on account of its size, since the circular orbits of the outer planets are much larger, but on account of its extraordinary usefulness in saving the apparent motions of not only the sun but also all the primary planets. ([2], VII, 403)

In these thoroughly informed and absolutely categorical affirmations by Kepler, Swerdlow sees only lack of knowledge and guesswork.

Swerdlow himself guesses:

> It is also possible that Regiomontanus saw no particular sense in assuming that an empty point, the mean sun, was the center of the planetary system. Kepler expressed his suspicion of this in *Mysterium cosmographicum*, cap. 18 . . . where he chose to measure planetary distances from the true sun. ([6], 472)

An unwary reader may perhaps suppose that Swerdlow's unspecified 'this' refers to Regiomontanus' attitude toward a central sun. But in the *Mysterium cosmographicum* Kepler refers only once to Regiomontanus. That unique reference has nothing to do with a central sun, but merely recalls Regiomontanus' (unfulfilled) intention to publish Ptolemy's *Music* ([2], I, 43; [5], 533).

Hence, Swerdlow's 'this' can refer only to Kepler's suspicion regarding an empty point as center. According to Swerdlow, 'Kepler expressed his suspicion of this in *Mysterium cosmographicum*, cap. 18 . . . where he chose to measure planetary distances from the true sun.' In Chapter 18 of his *Mysterium cosmographicum* Kepler expressed no such suspicion and made no such measurements.

Kepler wrote, let us recall, that 'the name applied by Copernicus to the true orbit of the earth about the sun' was *orbis magnus*. The projection of this curve on the celestial sphere is called the ecliptic and, as Copernicus said in the *Commentariolus*, 'the center of the earth always remains in the plane of the ecliptic' ([6], 445). An outer planet, however, crosses and recrosses the plane of the ecliptic from north to south and from south to north in what is known as its motion in latitude. The points where the planet's latitudinal motion cuts the plane of the ecliptic are the nodes. Copernicus' definition of these nodes is translated by Swerdlow as 'the intersections of the circles of the sphere and of the ecliptic' ([6], 482). But Copernicus' definition (*sectiones circulorum orbis et eclipticae*) concerned only two circles: 1) the ecliptic and 2) the planet's deferent, which he had just said remains in one plane with its epicycles (*circumferentiis quidem epicyclorum in una superficie permanentibus cum orbe suo*; 'the circumferences of the epicycles remain in one plane with their deferent'). There was no reference here to any sphere nor to any multiple 'circles of the sphere,' despite Swerdlow.

His translation imputes to Copernicus an ignorance of elementary plane geometry by having the astronomer say: 'the moon describes irregular circumferences of circles' ([6], 455). How can a circle's circumference be irregular? By definition, a circumference's regularity consists of the equidistance of each of its points from the center of the circle. Copernicus' expression (*inequales circulorum ambitus luna describat*) means 'the moon describes unequal peripheries of circles.' When two circles have unequal peripheries, their diameters are of unequal length, but each circle is perfectly regular. Copernicus' astronomy was introduced to Kepler by his teacher Michael Mästlin (1550–1631). In explaining Copernicus' lunar theory, Mästlin wrote:

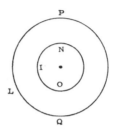

Mästlin's diagram for the moon (somewhat simplified)

For the new and full moon the epicycle of the apparent motion is NIO, but for the half moon PLQ ... The first epicycle becomes nearer to the earth and farther away ... because the apparent epicycle, as though composed of two, increases and decreases in size. ([2], I, 134)

Copernicus' 'peripheries of circles' (*circulorum ambitus*, *NIO* and *PLQ* in Mästlin's diagram) are unequal (*inaequales*). They are not 'irregular circumferences of circles,' despite Swerdlow.

When he encounters a similar statement in Copernicus' discussion of Mercury, Swerdlow shouts:

The statement that Mercury 'appears' to move in a smaller orbit when the earth is in the apsidal line and in a larger orbit when the earth is 90° from the apsidal line is utter nonsense as a description of the apparent motion of Mercury. No one ... gives such a description of Mercury's apparent motion. ([6], 504)

Such a description is given by Mästlin:

Whenever the earth is in the [apsidal] line of Mercury's apogee *B* or perigee *D* ... the circle of Mercury's path is *KML*, the smallest possible of all ... But if the earth, for instance at *C*, is midway between the apsides ... Mercury describes *PRQ*, the most extensive orb of its path. ... When the earth is in *B* ... Mercury's orb *KML* is the smallest and looks the smallest, because the earth is at its greatest distance from Mercury. With the earth in the perigee *D*, however, Mercury's orb is still minimal, but looks bigger because it is at its closest [to the earth]. When the earth at *C* is on the perpendicular [to the apsidal line], Mercury's orb is the largest, to be sure, but because it is at a greater distance from the earth, it looks no bigger. On the other hand, when the earth is at *S* ... according to Copernicus, the orb looks its biggest ... because that result is achieved in that position by the mutual balancing of distance and true size. ([2], I, 143)

Incidently (to use Swerdlow's idiosyncratic spelling, [6], 425, 430, 431, 433) Mästlin's description was translated into English on page 547 of the same issue that carries on page 504 Swerdlow's denial of the existence of any such description and on page 433 Swerdlow's citation of that translation.

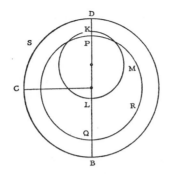

Mästlin's diagram for Mercury (somewhat simplified)

Swerdlow is so sure that 'No one ... gives such a description of Mercury's apparent motion' that he specifies: 'not even Copernicus in *De revolutionibus*' ([6], 504). In that work, however, Copernicus said: 'Mercury does not always describe the same circular circumference. On the contrary ... it traces an exceedingly varying circuit, smallest in point K, greatest in L, and mean in I' (V, 25); 'On account of this distance and its variation, the planet [Mercury] describes around F, the center of its orbit, unequal circles depending on the various distances' (V, 27); explaining why a certain angle has increased, Copernicus declared: 'the reason is ... that here the planet [Mercury] describes a larger circle than it does there' (V, 28).

In the *Commentariolus* Copernicus dissociated himself from the Pythagoreans because, in his judgment, they asserted the earth's motion gratuitously' (*temere mobilitatem telluris asseverasse*). As far as Copernicus knew, the Pythagorean astronomers did not provide powerful proof (*magnum ... argumentum*) of the earth's motion. Thus, they believed that the earth revolved, not around the sun, but around an unseen central fire, hidden from view by a supposed counter-earth. Swerdlow says: 'Copernicus ... accuses them [the Pythagoreans], justly I think, of not knowing what they were talking about' ([6], 439). Copernicus never accused the Pythagoreans 'of not knowing what they were talking about.' Swerdlow thinks that he did, and should have done so. Who does not know what he is talking about?

Swerdlow accuses Copernicus of failing to keep his promise to provide proof of the earth's motion in his exposition of the circles: 'Nowhere does Copernicus point out that any particular apparent motion . . . is evidence or proof of the earth's motion' ([6], 439).

Let us take a brief look at half a dozen passages in the *Commentariolus* where Copernicus attributed a 'particular apparent motion' to the real motion of the earth. In the first place, because the earth revolves annually around the sun at a distance which is negligible in comparison with the distance to the stars, 'it follows that the sun will appear to be carried around by this motion' ([6], 441). 'Further, the sun will appear to move nonuniformly in this motion on account of its distance from the center' of the earth's annual revolution ([6], 441). Third is the earth's 'daily rotation . . . on account of which the entire universe appears to be driven around' ([6], 444). Fourthly, the apparent precession of the equinoxes 'can occur through the movement of the earth' ([6], 445). Fifthly, a superior 'planet is seen sometimes to move retrograde and often to stand still, which does not result from the motion of the planet, but rather from the motion of the earth' ([6], 480). Lastly, 'Venus also is sometimes seen to move retrograde . . . In the case of the superior planets it happens because the motion of the earth exceeds their own motion, while in the case of Venus the motion of the earth is exceeded by the motion of Venus' ([6], 493). In these six instances, Copernicus pointed out, a particular apparent motion was evidence of the earth's motion. In this way Copernicus amply fulfilled his promise to provide 'considerable evidence' of the earth's motion 'in the explanation of the circles' ([6], 439).

The founder of modern astronomy did not exaggerate when he claimed that 'the motion of the earth by itself accounts for a considerable number of apparently irregular motions in the heavens' ([6], 436).

In his *Commentariolus* Copernicus presented the earliest substantial assertion of the heliocentric astronomy. In the opinion of the most recent translator of Copernicus' *Commentariolus*, it 'giv[es] (correctly) as the only evidence for his assertion the equivalence of heliocentric to geocentric planetary theory' ([6], 425). Nowhere in the *Commentariolus* (or anywhere else for that matter) did Copernicus ever equate heliocentrism with geocentrism. This non-Copernican, even anti-Copernican equation is the precise opposite of what Copernicus wrote over and over again concerning his relation to his geocentric and geostatic predecessors. Yet Swerdlow is not alone in proclaiming the correctness of this un-Copernican equation. In so doing, he ignores the difference between the points on some mathematical diagram and the real bodies in the actual universe. He neglects the crucial distinction between optical appearances and physical realities. He forgets about the flattening of the earth at the poles, the Foucault pendulum, annual stellar parallax, and the other momentous discoveries to which Copernicus' heliocentric planetary theory opened the way. In short, Swerdlow disregards the verdict of history: As long as the earth stood still, astronomy stood still.

BIBLIOGRAPHY

[1] JUSTINIAN, *Digest.*
[2] J. KEPLER, *Gesammelte Werke* (Munich, 1937).
[3] S. MÜNSTER, *Rudimenta mathematica* (Basel, 1551).
[4] P. NUNES, *Rules and Instruments of the Art of Navigation*, ch. 11, in *Opera* (Basel, 1566).
[5] J. REGIOMONTANUS, *Opera collectanea* (Osnabrück, 1972).
[6] N. SWERDLOW, 'The Derivation and First Draft of Copernicus's Planetary Theory – A Translation of the Commentariolus with Commentary,' *Proceedings of the American Philosophical Society*, CXVII (1973) 423–512.

14

Copernicus and his Relation to Italian Science

At the University of Bologna in 1496 the winter semester began on 19 October [1]. Among the newly enrolled students of canon law was a twenty-three year old dropout from Cracow University named Nicholas Copernicus (1473-1543). The official report of his matriculation at Bologna is missing, since " all of that university's documents prior to the sixteenth century have been completely lost [2] ". On the other hand, the " Records of the German Nation of Bologna University " have been preserved and published [3]. Eligibility for admission to this German Nation was based on language, not geography. In other words, a potential member's home did not have to lie within the boundaries of the Holy Roman Empire of the German Nation, but German had to be his *Muttersprache*. As this requirement was expressed by the 1497 statutes of the German Nation:

> We decree and ordain that in this beneficent city [Bologna] the students of canon or civil law who are of the nation of the Germans, that is, all those whose native language is German even though they may live elsewhere... shall be deemed and understood to be the Association of the German Nation [4]

In accordance with the terms of this statutory provision, Copernicus qualified for, membership. For, although his home was located in Poland, just outside the boundaries of the Holy Roman Empire of the German Nation, his native language was German. Like other German-speaking Polish students of law in Bologna Copernicus joined the German Nation.

At the time of his entrance into the German Nation a newly admitted member was expected to contribute a sum of money to help defray the expenses of the Nation's official ceremonies. No specified fee was required, each individual being permitted to determine the size of his own contribution. These varied considerably, amounting in Copernicus' case to nine *grossetti* [5]. The contemporary value of a *grossetto* may be estimated from the

---

(1) CARLO MALAGOLA, *Monografie storiche sullo Studio Bolognese* (Bologna, 1888), p. 379.

(2) GIORGIO CENCETTI, *Gli Archivi dello Studio Bolognese*, (Bologna: Zanichelli, 1938; Pubblicazioni del r. Archivio di Stato in Bologna, III), p. 66.

(3) *Acta nationis germanicae universitatis bononiensis*, edd. Ernst Friedländer and Carlo Malagola (Berlin, 1887).

(4) *Ibid.*, p. 4, lines 5-8.

(5) *Ibid.*, p. 248, line 43.

fact that one such small silver coin was the price of the Nation's palm branches for Palm Sunday in the jubilee year 1500 [6].

If Copernicus' nine *grossetti* do not seem to constitute a particularly generous contribution in comparison with the handsome donations offered by some of his fellow-entrants, it should be remembered that his nomination to a lucrative canonry was still being contested. That is why he prudently refrained from designating himself a canon at the time of his admission to the German Nation. About a year later, however, he received the welcome news that all opposition to his confirmation as a canon had at last been overcome. Hurrying with two witnesses to the office of a Bolognese notary, on 20 October 1497 he took possession by proxy of the canonry which he retained throughout the rest of his life [7]. While enjoying the ample income of his canonry, Copernicus steadfastly refused to become a priest, even when threatened on 4 February 1531 by his bishop with the loss of his emoluments unless he proceeded promptly to take higher orders and enter the priesthood [8].

The false assertion that Copernicus was ordained a Roman Catholic priest originated with Galileo, at a time when he was desperately striving by fair means or foul to save his church from making the deplorable mistake of condemning Copernicanism as heretical [9]. The farsighted wisdom of Galileo and the nearsighted obduracy of those who then occupied the seats of ecclesiastical power are demonstrated by the startling contrast between two actions on the part of the highest authorities: on 22 June 1633 Galileo, under prison sentence, was coerced into publicly proclaiming against his better judgment that the earth does not move [10]; some 330 years later, on 15 February 1964 the quadricentennial anniversary of the nativity of this quondam jailbird was celebrated in his birthplace Pisa by the president of the Italian Republic.

Half a century ago the spurious portrait of Copernicus as a priest was revived by a deliberate falsification of the aforementioned Bolognese notarial instrument. Whereas the notary, writing in Latin and employing a traditional legal formula, declared that Nicholas Copernicus had appeared before him in person (*personaliter constitutus*), the archivist who discovered

---

(6) *Ibid.*, p. 258, lines 6-7.

(7) LINO SIGHINOLFI, *Domenico Maria Novara e Nicolò Copernico allo Studio di Bologna*, « Studi e memorie per la storia dell'Università di Bologna, 5, 232–233 (1920).

(8) FRANZ HIPLER, *Nikolaus Kopernikus und Martin Luther*, « Zeitschrift für die Geschichte und Altertumskunde Ermlands », 1867–1869, 4, p. 502, n. 56, citing Bischöfliches Archiv zu Frauenburg (now Archiwum Diecezjalne w Olsztynie, that is, Diocesan Archive in Olsztyn), Al, fol. 264; HENRYK ZINS, *Kapituła fromborska w czasach Mikołaja Kopernika*, « Komunikaty mazursko-warmińskie », 66, 413 (1959); *Idem, Czy Mikołaj Kopernik miał święcenia kapłańskie?*, « Kwartalnik Historyczny », 68, 739 (1961).

(9) EDWARD ROSEN, *Galileo's Misstatements concerning Copernicus*, « Isis », 49, 319–320 (1958).

(10) ANTONIO FAVARO, *Galileo e l'inquisizione* (Florence, 1907), pp. 146–147.

and published this valuable paper pretended that *personaliter* was *presbiter*, thereby providing a fake documentary foundation for a resuscitation of Galileo's unfounded claim that Copernicus was a priest [11].

In 1496, although officially Copernicus had gone to Bologna, long famous as a center of legal instruction, to study canon law, his real love was astronomy. At Bologna University, " once matriculation had occurred, the student was free to attend courses, and the professors could not exclude him [12] ". Taking advantage of this liberal academic policy, Copernicus had the great good fortune to encounter in Bologna a highly competent astronomer, who had ingenuity enough to conceive new ideas and courage enough to publish them, even when they contradicted established authority. The name of that original thinker was Domenico Maria Novara (1454–1504).

As his surname " Novara " or " da Novara " indicates, that city in northwestern Italy had been the home of his ancestors. One of them, however, had been invited to move eastward across the top of the Italian boot to Ferrara, where our anti-traditionalist was born and christened Domenico Maria [13]. Hence he was variously known as Maria, Novara or da Novara, and Ferrariensis. In his own publications, however, he usually styled himself " Domenico Maria da Novara of Ferrara (Ferrariensis)," and described himself as holding two academic degrees, doctor of arts and also of medicine. It is not yet known when and where he pursued these studies.

An author named " Ferrariensis " was cited by Galileo in the student notes which he wrote in 1584 while he was enrolled at the University of Pisa. In the course of copying some still unidentified professor's commentary on Aristotle's *Heavens*, among those authorities who maintained that the universe can be eternal, Galileo listed this " Ferrariensis " [14]. Galileo's second citation of this same Ferrariensis [15] among those writers who thought that the heavens were composed of more than one element implies that this Ferrariensis was a follower of Thomas Aquinas and a commentator on his *Summa contra gentiles*. In that commentary Galileo's Ferrariensis attributed to Aristotle the opinion that celestial intelligences constitute the Heavenly Forms [16].

---

(11) EDWARD ROSEN, *Copernicus Was Not a Priest*, « Proceedings of the American Philosophical Society », *104*, 650–656 (1960); *Idem, Copernicus' Alleged Priesthood*, « Archiv für Reformationsgeschichte », *62*, 91–92 (1971) ; above, 48–49.

(12) CENCETTI, p. 67.

(13) LORENZO BAROTTI, *Memorie istoriche di letterati ferraresi* (Ferrara, 1792–1793), II, 26–27.

(14) *Le opere di Galileo Galilei*, national edition (Florence, 1890–1909; reprinted, 1968), I, *32*, 6. Galileo's Ferrariensis was a Dominican friar named Franciscus Silvester. According to Jacques Quétif and Jacques Echard, *Scriptores ordinis praedicatorum* (Paris, 1719–1723; reprinted by Burt Franklin, New York, 1959), II, 59–60, Franciscus Silvester's *Questions on Aristotle's Soul* and *Physics* were both published posthumously at Rome in 1577.

(15) GALILEO, « Opere », I, *76*, 33.

(16) *Ibid.*, I, *105*, 27. Franciscus Silvester Ferrariensis' commentary on Aquinas' *Gentiles* was first published posthumously at Paris in 1552.

In short, we discern in Galileo's three references to his Ferrariensis a scholastically oriented commentator on Aristotle's *Physics* and on Aquinas' *Contra gentiles*. Clearly, Galileo's Ferrariensis, a Thomist who died on 19 September 1528, was a very different person from Copernicus' Ferrariensis, an antitraditionalist practical astronomer, who died twenty-four years earlier [17].

Despite the marked differences between Galileo's Franciscus Silvester Ferrariensis (c. 1474–1528) and Copernicus' Domenico Maria Novara Ferrariensis (1454–1504), these two partly contemporary authors associated with Ferrara were merged into one by Tommaso Campanella (1568–1639). In his courageous defense of Galileo (*Apologia pro Galileo*, Frankfurt/Main, 1622), which Campanella wrote in 1616, while he was languishing in a Neapolitan dungeon without access to a library, the unfortunate prisoner said (Chapter II, no. 3, p. 9): " Prior to Copernicus, on the basis of observation of new phenomena Franciscus Maria of Ferrara taught that a new astronomy must be instituted, and this is what his disciple Copernicus did ". Mistakenly combining the (first) baptismal name of Galileo's Ferrariensis with the second baptismal name of Copernicus' Ferrariensis, Campanella emerged with an unhistorical " Franciscus Maria of Ferrara ". The historical Domenico Maria of Ferrara never " taught that a new astronomy must be instituted ", despite Campanella's retrojective inference from " what his disciple Copernicus did. " Copernicus' Bolognese teacher, Domenico Maria Novara, adhered to the fundamental doctrine of the old astronomy which held that the sun's motions were real, whereas in Copernicus' new astronomy those motions were downgraded to mere appearances. The epitaph on Novara's tombstone which was erected by Mariano Zuccati, one of his two heirs, solemnly declared that the deceased had been familiar with " the phases of the moon and the travels of the sun ". This last expression " travels of the sun " (*phoebique meatus*; Malagola, *Monografie*, p. 417) serves as a touchstone whereby the adherents of the old astronomy may be safely separated from the advocates of the new astronomy.

Copernicus' Ferrariensis, that is, Domenico Maria Novara, was a professor at Bologna University from 1483 to 1504. During each of those twenty-two years the subject of his lectures was officially designated " astronomy [18] ". Hence the description of Novara as a professor of astrology is a mistake [19]. The label " astrology " had been attached to the course prior to 1439, when

(17) ETTORE BORTOLOTTI, *La storia della matematica nella Università di Bologna* (Bologna, 1947), p. 20.

(18) UMBERTO DALLARI, ed., *I rotuli dei lettori legisti e artisti dello studio bolognese dal 1384 al 1799* (Bologna, 1888–1924) I, 121–185.

(19) ALBANO SORBELLI, *Il Tacuinus dell'Università di Bologna e le sue prime edizioni*, « Gutenberg Jahrbuch », *13*, 109 (1938); LYNN THORNDIKE, *A History of Magic and Experimental Science* (New York, 1923–1948), V, 234; ERNST ZINNER, *Entstehung und Ausbreitung der Coppernicanischen Lehre* (Erlangen, 1943), p. 161.

the name was changed to astronomy. This remained the standard nomen-clature thereafter, with the exception of 1443–1444 (" astrology "), and of six other years in which both " astrology " and " astronomy " were used (1448–49, 1463–64, 1468–69, 1470–71, 1473–75). During Novara's tenure, however, his course was invariably called " astronomy ". What the autho-rities thought of his performance may be judged from the increments they voted to add to his starting salary. In 1484 this was 50 Bolognese lire, raised to 100 in 1486, to 200 in 1498, and finally to 300 in 1502 [20].

In addition to his teaching duties, as professor of astronomy at Bolo-gna University Novara was required to compose for each year a prognosti-cation (*et fiat iudicium et tacuinum*) [21]. If Novara fulfilled this obligation faithfully, he would have prepared twenty-one such publications for the years beginning with 1484, following his appointment to the faculty of Bolo-gna University, and ending with 1504, the year of his death. However, these slender and ephemeral forecasts, of which only a relatively small number of copies was printed, have not in all cases survived. In fact, one speciali-zed student of this subject knew only twelve out of the presumed total of twentyone [22]. A recent investigator of *The University and the Press in Fif-teenth – Century Bologna* registers only nine, his earliest being that for 1487 and his latest that for 1501 [23]. This latter was written in Italian, being inten-ded for the townspeople of Bologna, whereas its Latin counterpart was aimed at the university population. Thus, Novara's last prognostication, that for 1504, was printed in two versions, Latin and Italian, which were once both owned by Prince Baldassarre Boncompagni, who published a bibliographical description of them [24]. One of the odd features of Novara's annual progno-stications is the frequent shift from one printer-publisher to another. All in all, at least six such investors in Novara's prescience are known.

Although the rest of Novara's prognostications soon passed into obli-vion, the prediction for 1489, published by Bazaliero Bazalieri, contained a striking passage which attracted the attention of a later professor of mathe-matics at Bologna University, who obtained his appointment in the face of no less doughty a competitor than the youthful Galileo [25]. The successful

(20) MALAGOLA, *Monografie*, pp. 441–443.

(21) DALLARI, I, 121.

(22) GUSTAV HELLMANN, *Versuch einer Geschichte der Wettervorhersage im XVI. Jahr-hundert*, « Abhandlungen der preussischen Akademie der Wissenschaften », physikalisch-mathematische Klasse, 1924, no. 1, p. 34.

(23) CURT F. BÜHLER, *The University and the Press in Fifteenth-Century Bologna* (No-tre Dame, Indiana, 1958; Texts and Studies in the History of Mediaeval Education, no. 7), pp. 65, 82, 83, 86, 92, 95, 96, 100.

(24) *Bullettino di bibliografia e di storia delle scienze matematiche e fisiche*, 4, 340–341 (1871), (reissued, New York: Johnson, 1964).

(25) ANTONIO FAVARO, *Galileo Galilei e lo Studio di Padova* (Padua; Antenore, 1956; reprint of Florence, 1883 edition), I, 19 (Contributi alla storia dell'Università di Padova, 3–4); MALAGOLA, *Monografie*, pp. 452–459.

rival, Giovanni Antonio Magini (1555–1617), in his *Tables of the Second Celestial Movables* (or Planets) said:

... Ptolemy's latitudes of places ought to be increased, in my opinion, not only for this reason [certain recent observations] but also on the authority of Domenico Maria of Ferrara. This man, endowed with a brilliant mind, was the teacher of Nicholas Copernicus. I propose to pass on to students Domenico Maria's opinion concerning this matter, especially because I know that it is not so easy for everybody to lay his hands on the writings of Domenico Maria. Now in a certain old prognostication for the year 1489, published at Bologna, Domenico Maria expresses himself as follows: « In recent years, when I was examining Ptolemy's *Geography*, I found the elevations of the North Pole, as given by him for individual regions, to be 1°10′ less than those in our time. This discrepancy cannot be ascribed to the defectiveness of [Ptolemy's] Table, for it is unbelievable that throughout its entire length the volume is uniformly deficient in the numerical values of the Tables. For this reason it is necessary to admit that the North Pole has shifted toward the zenith. Thus, continuous observation [over a period] of time has now begun to disclose to us what was concealed from our forerunners, through no fault of their own. On the contrary, the reason is that they lacked the observations over a long [period of] time of their own predecessors. For, the elevation of the pole had been observed before [the time of] Ptolemy in very few places, as Ptolemy himself indicates at the beginning of his *Geography*, where he states: [26] 'Hipparchus was the only one who transmitted to us the latitudes of a few places, whereas very many of the distances, especially those extending to the east or west, were established by a certain general tradition. This was not the fault of the authors themselves, but the reason was that there was as yet no practice in more accurate mathematics. ' No wonder that our predecessors did not perceive this extremely slow motion, for in 1070 years it discloses itself moved by scarcely 1° toward the zenith of the inhabitants. This is shown by the narrow Straits of Gibraltar, where in Ptolemy's time the elevation of the North Pole appeared to be 36¹/₄° above the horizon [Ptolemy's Mt. Calpe], but is now 37 2/5º. A similar discrepancy is revealed by Leucopetra [regarded by Ptolemy as the southern tip of Italy] in Calabria and by particular places in Italy, that is, those whose identification has not changed from Ptolemy's time to ours. Hence, as a result of this motion, those regions which are now inhabited will in the end become deserted places, whereas those which are now broiling in the Torrid Zone will be brought into our temperate region, although after a long passage of time since this very slow motion is completed in 395,000 years » [27].

Although Magini accepted Novara's secular shift in the direction of the earth's axis, this imaginary phenomenon was flatly rejected by Galileo. He did so in an undated marginal note on page 213 of his personal copy of William Gilbert's *Magnet* (London, 1600), where he wrote:

The incorrectness of Domenico Maria's observation is made particularly clear by the fact that if the elevation of the pole had become of this nature on our meridian, on others this would not have happened to the same extent; indeed, in still other regions the elevation would have changed practically not at all, and so on [28].

Galileo's reaction to Novara followed in the footsteps of Gilbert. But the English physician who discovered that the earth is a big round magnet

---

(26) This paraphrase of I, 4, was presumably based on the Bologna, 1477 edition of Ptolemy's « Cosmography », as it was entitled, sig. A2ᵛ. This edition was reprinted at Amsterdam in 1963 (Meridian Publishers).

(27) G. A. MAGINI, *Tabulae secundorum mobilium coelestium* (Venice, 1585), pp. 29-30.

(28) *Opere di Galileo*, VIII, 625, lines 1–4.

argued against Novara by citing contrary observations purporting to show that the latitude of the city where we are now assembled to discuss this matter had decreased by ½° since the time of Ptolemy, that is, from 41 2/3⁹ to 41 1/6°, according to Erasmus Reinhold [29].

A decade earlier, on 1/11 December 1590, the greatest astronomer between Copernicus and Galileo, the eminent Danish observer Tycho Brahe (1546–1601) wrote to Magini:

It occurs to me now that when I read the Introduction of your *Tables of the Second Movables* you shared the opinion of Copernicus ' teacher Domenico Maria of Ferrara, of most blessed memory, that the latitudes of places steadily change somewhat. In that work you quote from the same [Domenico Maria] the arguments which in your judgement can support that conclusion, whereas I believe that they are not strong and valid enough (I say this with the kind permission of both of you). Elsewhere I shall clearly demonstrate that this [axial shift] does not occur. In fact, the case of Rome all by itself proves that the polar elevations do not change. According to the observation of Regiomontanus [the greatest astronomer of the 15th century], Rome still has practically the same latitude as it formerly had in the time of Pliny [the Elder, who perished in the eruption of Mt. Vesuvius in 79 A. D.]. The investigation is carried out by means of the ratio of the vertical shaft to its shadow (provided that the proportion deals with the sun's center, not its uppermost limb). For in his *Natural History*, Book II, Chapter 72, Pliny says that in his time on the day of the equinox in the city of Rome one-ninth of the shaft was lacking in the shadow. Therefore, according to geometrical reasoning, at that time the sun's altitude at noon was 48°22' at the top of the limb. Hence, after subtraction of its apparent radius of 16', the altitude of its center comes out 48°6', also showing the angular distance of the equator to the same place. Nor do refraction or solar parallax make any perceptible difference in this instance, since they virtually cancel each other out. As a result, Rome's latitude emerges from this investigation as 41°54' [+48°6'=90°]. I shall compare this result in the following way with the observation made at Rome by Regiomontanus in the period before ours [2 and 11 January 1462: 42°7'0'', 42°0'10''] . . . Hence, Regiomontanus' later observation, which seems more suitable than the earlier one, yields a latitude for the city of Rome around this time of almost exactly 42°. For the age of Pliny, from the shadow of the shaft it was found smaller by only six almost imperceptible minutes. Therefore it is accordingly quite clear that after the passage of so long a time the latitude has undergone no perceptible change. [Brahe then analyzes the comparable data for Venice and Ancona] . . . In the cases of these three cities of Italy which keep their latitudes unchanged for so many centuries, it is overabundantly clear that the elevations of the poles do not increase even when very many years intervene nor are they subject to any alteration. For, a discrepancy of 1 1/6° (the amount claimed by Domenico Maria) could not escape detection in this analysis . . Nevertheless, I should like to look at the [printed] books of that most excellent Maria, if any are available, for I have never seen them, or even at the manuscripts if they are preserved anywhere [30].

(29) WILLIAM GILBERT, *De magnete*, VI, 3 (London, 1600), pp.212–213. In Ptolemy's *Geography*, III, 1, Rome's latitude was given as 41°40' (1477 ed., sig. Cl ᵛ). But in Reinhold's *Prussian Tables* (2nd ed., Wittenberg, 1585), Catalogue of Certain Important Places, p. 7, Rome's latitude was given as 41°45', not 41°10'. Today the latitude of Monte Mario Observatory in Rome is given as about 41°55½'.

(30) *Tychonis Brahe dani Opera omnia* (Copenhagen, 1913–1929), VII, 297, 8–298, 40. Nearly three decades later, in a letter sent from Venice on 14 March 1619 Vincenzo Bianchi informed Johannes Kepler that the « writings of this eminent mathematician [Novara] are in the possession of my cousin, Abbot Lodovico Marcello »; KEPLER, *Gesammelte Werke*, XVII (Munich, 1955), *339*, 11–12.

The foregoing reactions to Novara's claim that the earth's northern latitudes increase slightly in the course of centuries are an excellent illustration of how science progresses as a self-correcting, multi-national enterprise. Magini, who was named to his professorship at Bologna University on 4 August 1588 [31], was presumably affected by local pride when he approved the conclusion of Novara, whose professorship at the same institution had commenced a little more than a century earlier. On the other hand, Novara's reasoning was flatly rejected by Brahe, Gilbert, Galileo, Willebrord Snel [32], Giovanni Battista Riccioli [33], and others.

Gilbert's example affected the attitude of Johannes Kepler (1571–1630) toward Novara. The discoverer of the three principles of planetary motion had examined sailors' reports concerning the declination of the compass needle from the geographic north pole. From the data available to him, as he interpreted this information, Kepler drew the conclusion that " there is a distance of 6 1/2° between the magnetic pole and the celestial pole " [34] (or the point where the prolongation of the earth's axis passing through the terrestrial north geographic pole intersects the heavenly vault). As a pious Christian who believed that the world owed its origin to a divine Creator, Kepler reasoned that

when the universe began, the point toward which the magnet is oriented was the terrestrial pole. Since that time the earth's pole has shifted from its original position through 6¹/₂° along the arc of a great circle from the Anian [Bering] Strait toward the Azores. This [motion] is confirmed by Antonio [sic] Maria, who states that from Ptolemy to us the pole has approached us by 1°10′. This uniform shift he finds in very many places in Italy and at Cadiz. Add Pliny's shadow staff. Had Maria remarked that the pole moves away from us through an equal distance, he would have said the opposite of my conjecture based on the magnet. But because he states that the pole approaches Italy, he is on my side. For a great circle drawn at right angles to the meridian of the Azores keeps Italy in the middle of the Azores. The computation is also in agreement. For if 1072 years make a degree, then 5600 years have produced about 5 1/5° [35].

Kepler reckons with 5600 years, which he believed to be the age of the world in his own time. His use of 1072 years, instead of Novara's 1070, while a trivial difference, indicates that he did not have Novara's text itself before

(31) MALAGOLA, *Monografie.* p. 458; FAVARO, *Galileo... e Padova*, I, 19.

(32) SNELLIUS, *Eratosthenes Batavus* (Leiden, 1617), p. 40.

(33) RICCIOLI, *Almagestum novum* (Bologna, 1651), II, 348–349. Riccioli mistakenly classified Giordano Bruno, together with Magini, as a supporter of Novara's view. As the foundation of this classification, Riccioli cited Bruno's *De maximo et immenso* (Frankfurt Main, 1591), p. 306 (Book III, Chapter 5). But Bruno's reference there to a motion of the earth's pole has no connection whatever with the idea of Novara, whose name he does not mention. Indeed, there is no firm evidence that Bruno ever heard of Novara.

(34) JOHANNES KEPLER, *Gesammelte Werke*, XIII (Munich, 1945), *351*, 455–456, in a letter of 30 May 1599 from Kepler to Herwart von Hohenburg, chancellor of Bavaria.

(35) *Ibid.*, XIV, *26*, 188–198, in a letter of 6 August 1599 from Kepler to Herwart. In a letter of 19/29 August 1599 to Michael Mästlin, his former professor of astronomy at Tübingen University, Kepler's reasoning is much the same (*ibid.*, XIV, *55*, 514–522).

him, as does also his use of Antonio, instead of Domenico, as Novara's first baptismal name. Moreover, Novara did not draw Kepler's distinction between two north poles, one geographic and the other magnetic, joined together at the time of the supposed creation, and then drifting slowly apart for thousands of years. The motion of Novara's unique north pole was such that latitudes in Italy had *in*creased by 1°10′, but Kepler misunderstood "Maria to say that the latitudes in Italy had *de*creased by 1°10′ from Ptolemy ". Kepler included this misunderstanding in a manuscript discussion of the earth's motions that was first published long after his death [36]. This undated manuscript must have been written before 12 January 1603, when Kepler states that his reading of Gilbert caused the

complete evaporation of my speculations concerning that slow shift of the pole of the daily motion away from [those] places on the earth that had been located below it at [the time of] creation. [37]. I set this speculation aside all the more gladly because I find that here in Prague the elevation of the pole as observed 200 years ago agrees quite closely with today's value [38].

The rector of Prague University had called Brahe's attention in 1600 to a manuscript whence Kepler, who was then Brahe's assistant at Prague, copied out an excerpt [39], which he reported as follows:

Here at Prague 200 years ago [1416, according to the excerpt] a certain Doctor Syndel observed the sun's noon altitude on an equinoctial day and [the day of the] summer solstice. From these values (after subtraction of the sun's place, according to Tycho), the altitude of the pole emerges as 50°4′20″ [50° 5′22″, according to the excerpt, in striking agreement with the modern value, 50°5′19″] . . . as it is found today [40].

This stability of the latitude of Prague was cited by Kepler when he categorically denied Novara's thesis in Book III of his *Epitome of Copernican Astronomy* (Linz, 1618). Using the catechetical format, Kepler posed the following Question:

By what evidence is it clear that the elevation of the pole at places on the earth's surface is always the same?

[Answer]: The elevation of the pole at Prague was observed 200 years ago to be 50°6′, just as it is today.

[Question]: But do we read that 100 years ago Joh. [sic] Maria had his doubts about this, as the result of a comparison of the Ptolemaic geography with the modern?

(36) *Joannis Kepleri opera omnia*, ed. Christian Frisch (Erlangen Tübingen, 1858–1871), VIII, 235, 25-26.

(37) EDWARD ROSEN, *Kepler's Somnium* (Madison, University of Wisconsin Press, 1967), pp. 98–99. This book was awarded the Charles Pfizer Prize in 1968 by the History of Science Society, USA.

(38) KEPLER, *Gesammelte Werke*, XIV, 347, 213-217, in a letter of 12 January 1603 to Herwart.

(39) BRAHE, *Opera*, V, 228.

(40) KEPLER, *Gesammelte Werke*, XIV, 208, 232-235, in a letter of 10/20 December 1601 to Mästlin.

[Answer]: The fault is believed to attach to Ptolemy. He did not make observations personally in the places of the west, but undoubtedly copied these values from an inaccurate geographical table or inferred them from the length of the summer [solstitial] day, as he had learned to do from a crude prescription [41].

Thus we see that after an initial acceptance of Novara, Kepler reversed his attitude and completely rejected the Bolognese professor. But Kepler's early approval of Novara was based on an imperfect understanding of Domenico Maria's point of view. The same cannot be said about Copernicus, who told his only disciple, George Joachim Rheticus (1514–1574), that he had been "not so much the pupil as the assistant and witness of observations [42]" of Novara. In that capacity Copernicus "had lived with Domenico Maria of Bologna, whose thinking he had mastered thoroughly [43]".

Hence it was on the basis of complete information that Copernicus, like Kepler in the *Epitome of Copernican Astronomy*, rejected Novara's thesis outright in the *Revolutions*, II, 6:

The altitudes of the pole, or the latitudes of the places, and the shadows on the equinoctial days agree with the recorded ancient observations. This had to happen, because the equator follows the pole of the terrestrial globe [44].

With his characteristic courtesy, Copernicus refrains from mentioning here the name of his teacher Novara. Copernicus' silence has been misinterpreted by a spiteful maligner of the Polish astronomer to show that he was ungrateful to those from whom he learned [45]. But Copernicus did cite Novara by name in his discussion of various determinations of the obliquity of the ecliptic [46]. True, in the final draft of his *Revolutions* (III, 6) for the printer, Copernicus dropped Novara, while leaving two much better known fifteenth-century astronomers, Peurbach and Regiomontanus. But his reference to Novara in his autograph proves that he had no intention of concealing the name of his teacher at Bologna.

According to the French biographer of Copernicus, Pierre Gassend (1592–1655), Novara "was greatly delighted that Copernicus did not reject the idea which gripped him that the elevation of the pole in the same place was not

(41) *Ibid.*, VII, 147, 9–17.

(42) EDWARD ROSEN, *Three Copernican Treatises*, 3rd ed. (New York: Octagon, 1971), p. 111.

(43) RHETICUS, *Ephemerides novae ... ad annum MDLI* (Leipzig, 1550), preface; reprinted in Leopold Prowe, *Nicolaus Coppernicus* (Osnabrück, Zeller, 1967; reissue of Berlin, 1883-84 edition), II, 390. Novara could more readily receive Copernicus in his own home since he was a childless bachelor and no servants are mentioned in the notarial inventory of his worldly goods (Sighinolfi, pp. 213, 235).

(44) NICHOLAS COPERNICUS, *Complete Works*, II (London, MacMillan, 1974), English translation of the *Revolutions* by Edward Rosen.

(45) ARTHUR KOESTLER, *The Sleepwalkers* (New York, 1958, 1968), p. 205.

(46) COPERNICUS, *Complete Works*, I, (London, MacMillan, 1972), fol. 79ʳ.

constant as was commonly believed " [47]. Gassend's " greatly delighted " (*Delectavit autem illum* [that is, Novara] *maxime*) was erroneously transferred from Novara to Copernicus by Malagola. In his biography of Antonio Urceo, the professor of Greek at Bologna University when Copernicus was a student there, Malagola declared that " Copernicus was greatly delighted with this discovery [by Novara], as Gassend said " [48]. But according to Gassend's (mistaken) statement, it was Novara who " was greatly delighted that Copernicus did not reject his idea " (... *non improbari Copernico suspicionem* ...). Malagola was misled by Domenico Berti, who also erroneously identified Novara's shift of the earth's axis away from the equator, thereby increasing northern latitudes, with the precessional movement, in which the earth's axis is locked perpendicularly to the equator, so that terrestrial latitudes are not affected [49].

Gassend's misstatement that Copernicus did not reject Novara's increased latitudes is by no means the only error in Gassend's biography. Nevertheless, it would be wrong to conclude, because Copernicus discarded Novara's supposed discovery, that the Bolognese professor had no influence on his Polish student and assistant. For even if Novara was mistaken, he did have the courage to assert in a printed work that the earth was not completely immovable. In so doing, he implicitly contradicted a fundamental doctrine of the Aristotelian philosophy and Ptolemaic astronomy. Of all of Copernicus' numerous scientific achievements, the greatest unquestionably was his recognition of the true cosmic status of the earth as a natural satellite of the sun in unceasing motion. It is hard to believe that his close contact with Novara and his awareness that, according to Novara, the earth's axis shifted did not in some way and to some extent contribute to Copernicus' concept of the moving earth, which is the principal reason for our commemorative meeting here today.

(47) GASSEND, *Vita* ... *Copernici* (Paris, 1654), pp. 5–6; *Opera* (Florence, 1727), V, 441: *Delectavit autem illum [that is, Novara] maxime non improbari Copernico suspicionem, qua tenebatur, ne Poli in eodem loco altitudo non tam constans foret, quam vulgo haberetur.* The Lyon, 1658 edition of Gassend's *Opera* has been reprinted (Stuttgart-Bad Cannstadt, 1964).

(48) MALAGOLA, *Della vita e delle opere di Antonio Urceo* (Bologna, 1878), p. 349; *idem, Monografie*, p. 414.

(49) DOMENICO BERTI, *Copernico e le vicende del sistema copernicano in Italia* (Rome, 1876), pp. 39–41.

15

Nicholas Copernicus and Giorgio Valla

To understand the intellectual development of those thinkers who have helped to shape mankind's present view of the physical world, the books they consulted may offer useful clues. In the case of Nicholas Copernicus (1473-1543), the founder of modern astronomy, the effort to identify his working library has focused mainly on artifacts. Books have been preserved which were presented to him, or bequeathed by him, or marked as his property by his signature as owner, or annotated by him although they belonged to the Cathedral Chapter of Frombork, of which he was a canon.[1] But besides this external evidence of extant tomes and an inventory of his Chapter's library, internal evidence is provided by his unacknowledged borrowing from earlier or contemporary authors.

Giorgio Valla (1447-1500), for instance, published many works,

[1] Paweł Czartoryski, « The Library of Copernicus », *Science and History* (Studia Copernicana, XVI), Wrocław 1978, pp. 355-396.

none of which was owned by Copernicus,[2] whose writings never mentioned Valla's name. Yet Copernicus used Valla's posthumous *De expetendis et fugiendis rebus* (Venice, 1501), that vast encyclopedia in two volumes, whose title may be equated with « What to Seek and Shun ». « I shall shun » it, said a critic,[3] while « I shall seek good books ». Copernicus both sought it and shunned it.

Thus, in his *Revolutions* (*De revolutionibus orbium coelestium*, Nuremberg, 1543), I, 3, Copernicus correctly attributed to the early Greek thinker Xenophanes the conception of the earth's « under side descending endlessly » (*ex inferna parte infinita radicitus*).[4] Copernicus' immediate source unquestionably was *ex inferna parte in infinitum ... radicitus* in Valla's *Seek and Shun*.[5] Thus far Copernicus sought Valla but soon had to shun him, since his full statement about the shape of Xenophanes' earth was: *ex inferna parte in infinitum crassitudine radicitus immissam esse*. This *crassitudine* (thickness) must have discomfited Copernicus because it links up with nothing else in Valla's quoted statement. To forge such a link, Copernicus altered Valla's *immissam* (planted) to *submissa* (diminished), emerging with *crassitudine submissa* (diminishing thickness). Consequently, for Copernicus, the under side of Xenophanes' earth descends endlessly « with diminishing thickness » like an inverted cone. Xenophanes' conoidal earth is found only[6] in Copernicus, and in him only because he felt he had to shun Valla.

Copernicus knew that others before him had regarded the earth as a body in motion. To show that, far from introducing a novel idea, he was recalling an old geokinetic doctrine, he searched for his ancient predecessors, and named as many as he could find. Yet his *Revolutions*, as first printed, although misattributing to Aristarchus of Samos two views he never held – misattributions due to Copernicus' erratic source, the Venice 1515 Latin translation of Ptolemy's *Syn-*

---

[2] When the Frombork Cathedral library was catalogued on 22 September 1598, the philosophy section included « Georg. Valla », bound in red leather, with no indication of the title. See FRANZ HIPLER, *Analecta Warmiensia*, « Zeitschrift für die Geschichte und Alterthumskunde Ermlands », 5, 1870-1874, pp. 375/4↑.

[3] *Menagiana*, 3rd ed. (Paris 1715), II, 326/9: *Libros ego bonos expetam, fugiam tuum.*

[4] *Nicholas Copernicus Complete Works* (cited hereafter as NCCW), I (Warsaw/London 1972), fol. 3r/16; II (Baltimore 1978), 10/23-24.

[5] XXI, 41; sig. kk 7v.

[6] MARIO UNTERSTEINER, *Senofane* (Florence 1956), pp. 86-91, # 47; pp. 138-139, # 28.

taxis[7] – did not associate Aristarchus with the other geokineticists. But the recovery of Copernicus' autograph draft of the *Revolutions* disclosed a deleted reference to Aristarchus' geokineticism in Copernicus' manuscript. This deleted passage was absent from the first four editions of the *Revolutions* (1543, 1566, 1617, 1854), but was reinstated in all the later editions (1873, 1949, 1972). After alluding to an ancient Pythagorean's awareness of the earth's mobility, the deleted passage added: « some people say that Aristarchus of Samos was also of the same opinion » (*etiam nonnulli Aristarchum samium ferunt in eadem fuisse sententia*).[8]

With regard to Aristarchus' geokineticism, Copernicus had excellent reasons for concealing his source behind his deliberately vague expression « some people say » (*nonnulli ... ferunt*). For when Valla declared[9] that Aristarchus' « earth moves around the solar circle » (*terram moveri circa solarem circulum*), he left that last term unexplained. But when combined with Valla's immediately preceding statement that « Aristarchus makes the sun stationary » (*Aristarchus ... solem locat*), it could mean that Aristarchus' earth moved around the circle that had been the pre-Aristarchan sun's orbit. So far so good, since that was essentially what Copernicus himself did: his earth moved around the circle previously assigned to the sun, which he made stationary. But Valla has Aristarchus put his stationary sun « beyond the fixed stars » (*Aristarchus post non vagas stellas solem locat*). Why should Copernicus, a practical astronomer, call attention to an earth pursuing its orbit beyond the stars? No wonder Copernicus shunned Valla, and resorted to « some people say » instead.

These unacknowledged borrowings from Valla were used by Copernicus in *Revolutions*, I: Chapter 3 for Xenophanes, and for Aristarchus the deleted passage following Chapter 11. These sections were written by Copernicus at an early stage in his composition of the *Revolutions*.[10] At that time he was still unaware that XXI, 24, 41, in Valla's *Seek and Shun* were translations or paraphrases of an ancient Greek

---

[7] EDWARD ROSEN, *Aristarchus of Samos and Copernicus*, « Bulletin of the American Society of Papyrologists », *15*, 1978, pp. 85-93 ; above, 7-8.

[8] NCCW, I, fol. 11*v*/15-16; II, 25/21-22.

[9] *Seek and Shun*, XXI, 24; sig. kk 6*v*. This passage was quoted by LUDWIK ANTONI BIRKENMAJER, *Stromata Copernicana* (Cracow 1924), p. 163, in a discussion of Copernicus' indebtedness to Valla.

[10] EDWARD ROSEN, *When Did Copernicus Write the Revolutions?*, « Sudhoffs Archiv », *61*, 1977, pp. 144-155, here 150-153; above, 111–14.

author. Although in a private letter of 19 July 1498 Valla told a correspondent that in his *Seek and Shun* « you will see neglected ...
and bu-
ried [material] taken from the most outstanding Greek writers »,[11]
in *Seek and Shun* itself Valla generally avoided making any such ac-
knowledgment. In June 1542,[12] however, when Copernicus wrote
the preface to the *Revolutions*, he quoted from the Greek original which
Valla had covertly transformed into Books XXI-XXII of his *Seek
and Shun*. By the time Copernicus in 1542 quoted lines 17-21 on page
328 [13] in *Plutarchi opuscula LXXXXII* (Venice, 1509), he may have
learned that this (pseudo-Plutarchan) Greek work underlay Valla's
Seek and Shun, XXI-XXII. But decades earlier, when he was composing Book I of the *Revolutions*, he gave no sign of such awareness.

Even earlier, when Copernicus wrote his *Commentariolus*, he relied on the translation of Cleomedes by Valla in his composite volume
Georgio Valla Placentino Interprete, Hoc in volumine haec continentur:
Nicephori logica ... (Venice, 1498).[14] For in the *Commentariolus* Mars
completes its revolution in the 29th month. (*29. mense*),[15] a mistake
previously made only by Cleomedes (I, 3).[16] However, in his mature

---

[11] J. L. HEIBERG, *Beiträge zur Geschichte Georg Valla's und seiner Bibliothek*, « Centralblatt
für Bibliothekswesen », « Beiheft », *16*, 1896, p. 90/11-12.

[12] P. 146 at. n. 9, in the article cited in n. 10, above.

[13] Not 313, as on p. 90/8↑ in the article cited in n. 7, above.

[14] Valla's composite volume was issued in Venice on 30 September 1498 by Simon
Bevilacqua of Pavia; see *Indice generale degli incunaboli delle biblioteche d'Italia*, IV (Rome
1965), pp. 127-128, # 6792. Bevilacqua's book has been mistakenly described as a reprint
of a composite volume published a decade earlier in Venice by Antonio de Strata of Cre-
mona; see *Gesamtkatalog der Wiegendrucke*, I (Leipzig 1925), pp. 426-427, # 860. Strata's
publication began with Valla's translation of Alexander of Aphrodisias' *Problems*, printed
on 24 November 1488, and issued several weeks later in combination with two other
works. The presence of Valla's translation of Alexander of Aphrodisias' *Causes of Fevers*
in Bevilacqua's volume may have been responsible for the misdescription of Bevilacqua's
1498 publication as a reprint of Strata's 1488 publication. Actually, none of the three
components of Strata's 1488 volume reappeared in Bevilacqua's 1498 publication. Yet
the reprint misconception is traceable from G. D. Mansi's edition of J. A. Fabricius'
Bibliotheca latina mediae et infimae aetatis, VI (Padua 1754; reprint, Florence 1859), p. 572 /
right/ # 2, to T. L. HEATH, *Aristarchus of Samos* (Oxford, 1913; reprint, 1959), p. 321/9-12,
to G. SARTON, *Introduction to the History of Science*, I (Baltimore 1927), p. 157/8, and to
GRANT MCCOLLEY, GEORGE VALLA: *An Unnoted Advocate of the Geo-Heliocentric Theory*,
«Isis», *33*, 1941-1942, p. 314/5-10. In Bevilacqua's 1498 volume Valla is named as translator
over and over again, by contrast with *Seek and Shun*.

[15] EDWARD ROSEN, *Three Copernican Treatises*, 3rd ed. (New York 1971; cited hereafter
as « TCT »), p. 74/11.

[16] *Valla Interprete*, sig. i1v/15: *duobus annis & mensibus quinque*; Cleomedes, *De motu
circulari corporum caelestium libri duo*, ed. Hermann Ziegler (Leipzig 1891), p. 30/25.

work, the *Revolutions*, Copernicus discarded Cleomedes' misinformation and reduced Mars' period to two years,[17] closer to the correct number, 687 days or about 22½ months.

This is the sidereal period of Mars revolving around the sun, as it would be computed by an imaginary observer on the sun watching Mars leave the longitude of a given star and returning to it. But when Mars' relation to the sun is observed from the earth, it returns to the same position in its synodic period of 780 days (or 26 months). This number is given correctly for Mars by Cleomedes in a later section (II, 7),[18] where he lists the synodic periods for the five then known planets, as taken from a different source in close agreement with Ptolemy. Cleomedes' sources are unidentified or lost. His 29-month sidereal period for Mars, we are told,[19] « can hardly be anything but a simple mistake committed by some early compilator ». But Geminus gave Mars a 30-month period, « two years and six months, with two months and a half for a zodiacal sign »[20] ($2\frac{1}{2}^m \times 12 = 30^m = 2^y6^m$). Are we to invoke another simple mistake of 30 months for Geminus alongside Cleomedes' 29-month mistake? With regard to Cleomedes' supposed repetition of a previous compiler's simple mistake of 29 months for Mars' *sidereal* period, we are also told,[21] « The correct value must be very near to 2 years 50 days, the period given in the subsequent list » of *synodic* periods by Cleomedes (II, 7). But would the (hypothetical) compiler have confused Mars' sidereal and synodic periods? He could hardly have confused ἡμερῶν ψπ′ (780 days) with διετία καὶ πέντε μησί[22] (2 years and 5 months).

Copernicus never saw Cleomedes' Greek text, which was printed for the first time in Paris in 1539,[23] four years before he died in faraway Frombork. Nor did he know about Carlo Valgulio's translation of Cleomedes into Latin, published in Brescia on 3 April 1497.[24] This

[17] NCCW, I, fol. 9*v*/3↑ and diagram: *Martis bima revolutio*; II, 21/3↑.

[18] *Valla Interprete*, sig. l 3r/6↑: *diebus septingentis octoginta*; Ziegler p. 226/22.

[19] OTTO NEUGEBAUER, *A History of Ancient Mathematical Astronomy* (Berlin/New York 1975), p. 964/D/5-6.

[20] GEMINUS, *Elementa astronomiae*, ed. Karl Manitius (Leipzig 1898), p. 12/17-18.

[21] NEUGEBAUER, p. 964, n. 3.

[22] ZIEGLER, p. 30/25, p. 226/22.

[23] Facsimile of the title page in GEORGE SARTON, *A History of Science: Hellenistic Science and Culture* (Cambridge, MA, 1959), p. 306, fig. 56.

[24] *Gesamtkatalog der Wiegendrucke*, VI (Leipzig 1934), p. 729; facsimile of the title page in SARTON, p. 306, fig. 55.

marked the first appearance of Cleomedes in print, a year and a half before Valla's translation, which has been mislabeled Cleomedes' *editio princeps* when Bevilacqua's 1498 volume is misrepresented as a reprint of Strata's 1488 volume.

Mars' mistaken 29-month period was found by Copernicus in the Latin translation made from a Cleomedes Greek manuscript by Valla, whose extensive collection was « bought from the heirs of Giorgio Valla ... for the price of 800 gold » scudi by Alberto Pio of Carpi (c. 1475-1531). This prince bequeathed his own collection to his nephew Rodolfo Pio,[25] who was deeply in debt when he died on 2 May 1564. Two weeks later, in order to pay off this cardinal's clamorous creditors, an inventory of his library was begun to entice prospective purchasers. The copy sent to the eventual buyer, Alfonso II, duke of Ferrara, survives in Modena. For, nearly a quarter-century after his purchase of the Rodolfo Pio collection in February 1573, Alfonso II died on 27 October 1597, precipitating the passing by devolution of the duchy of Ferrara to the papacy, which agreed to the removal of the ducal collection to Modena.[26]

Two copies of the 16 May 1564 inventory of the library of Rodolfo Pio survive. The one in the Vatican Library,[27] published by Heiberg in 1896,[28] lacks the index present in the Modena copy. In the collation of both copies, as published by Giovanni Mercati in 1938,[29] the index lists not only the Greek text of Cleomedes but also a translation, Valla's.[30] On fol. 44*v* the text has remarks by Valla in Greek, and on fol. 1*v* the indication by Alberto Pio's teacher of Greek that the text belonged to his pupil.[31] Thus, Modena preserves Valla's Greek manuscript of Cleomedes together with his translation, sold

---

[25] GIROLAMO TIRABOSCHI, *Biblioteca modenese*, IV (Modena 1783), 163/9-12, 10↑-6↑.

[26] DOMENICO FAVA, *La Biblioteca Estense nel suo sviluppo storico* (Modena 1925), pp. 155, 164-165.

[27] Barberini latin. 3108 (formerly XXXIX, 12) fol. 55*v*-62*r*.

[28] At pp. 118-126 in the work cited in n. 11, above.

[29] *Studi e testi*, # 75 (Vatican City, 1938), pp. 223-241.

[30] *Ibid.*, p. 235/C/6-7.

[31] THOMAS WILLIAM ALLEN, *Notes on Greek Manuscripts in Italian Libraries* (London 1890), p.20, # 215; VITTORIO PUNTONI, *Indice dei codici greci della Biblioteca Estense di Modena*, « Studi italiani di filologia classica », 4 (1896), pp. 510-511, # 215. This number corresponds to # 65 in the inventory of Alberto Pio's collection, and to # 88 in Mercati (pp. 212, 229, 245). A second copy of Cleomedes was received by Alberto Pio as a gift, not by purchase (Fava, p. 150-5).

to Alberto Pio, bequeathed to Rodólfo Pio, bought by Alfonso II, and transferred to the Biblioteca Estense.

Some uncharitable people today may blame Copernicus for repeating in his *Commentariolus* the mistaken period for Mars that he found in Valla's translation of Cleomedes. On the other hand, Copernicus shunned the erroneous part of the reference to Aristarchus in Valla's *Seek and Shun*. Copernicus may have learned only later (if at all) that this reference occurred, not in an original work by Valla, but rather in Valla's translation or paraphrase of (pseudo-) Plutarch's *Opinions of the Philosophers*. Yet four hundred years later, in our own century, McColley asserted that Valla « set forth ... the geo-heliocentric theory »[32] in *Seek and Shun*, XVIII, 1. McColley was utterly unaware that *Seek and Shun*, XVIII, incorporates Valla's translation or paraphrase of Proclus' *Hypotyposis*.[33] In *Seek and Shun*, XVIII, it is Proclus who speaks to us through Valla's Latin rendering. In *Seek and Shun*, XVIII, Valla advocates nothing.

McColley applies the term « geoheliocentric » to the theory allegedly set forth by Valla:

Geocentric in respect to the seven « planets », Saturn, Jupiter, Mars, the sun, Mercury, and Venus [a total of six, be it observed, since McColley omitted the moon]; and heliocentric to both the true superior and the inferior planets, this system is aptly termed the geo-heliocentric.[34]

The system under discussion by Proclus, speaking through Valla, regarded Mercury and Venus (McColley's « inferior » planets) as revolving about the sun. To this extent, this system is heliocentric. But Mercury and Venus, revolving about the sun, are sometimes below it or « inferior », and at other times above it or « superior ». They accompany the sun, which in this system revolves around a stationary earth, and the periods of all three are equal.[35] Hence, an accompanying planet's circumsolar revolution resembles an epicycle, but no epicycle is allotted by this system to Mercury or Venus. Yet, according

---

[32] P. 314/11 in the article cited near the end of n. 14, above.

[33] *Procli Diadochi Hypotyposis astronomicarum positionum*, ed. Karl Manitius (Leipzig 1909), p. v/7-13.

[34] McColley, p. 313/10-13.

[35] *Hypotyposis*, ed. Manitius, p. 12/25-26; *Seek and Shun*, XVIII, 1; sig. ee6v/16↑: *Solis autem et Mercurii et Veneris aequales esse cursus intuemur.*

to McColley, « the sun carried their epicycle about the earth ».[36] Proclus-Valla's sun carried nothing. Proclus-Valla's Mercury and Venus
had their own independent motion apart from any epicycle.

The superior or outer planets, « Saturn, Jupiter, and Mars, revolved about both sun and earth », according to McColley's account
of this system.[37] « The earth also stood in the center of the universe »,[38]
around which the three outer planets revolved geocentrically. In so
doing, within their own orbits they completely enclosed the sun's
geocentric orbit, without intersecting it anywhere. Thus they may be
said to be revolving outside the moving sun, but not about it. For,
the sun was not the center of their orbits and, like them, it revolved
around the stationary earth, which was the center of the orbits of the
sun and of the outer planets. To say with regard to the three outer
planets, as McColley did, that they « revolved about both sun and
earth » suggests that they were both heliocentric and geocentric. The
outer planets were geocentric, and only the planets accompanyng
the sun were heliocentric. As Proclus says, speaking through Valla:

Of these five planets, some may be at any elongation from the sun, dia-
metrically [opposite it, or 180º away], sometimes in the trinal configuration
[at two vertices of an equilateral triangle inscribed in a circle, or 120º away],
at other times in the square configuration [at two adjacent vertices of a square
inscribed in a circle, or 90º away], and at still other times in the hexagonal
configuration [or 60º away], for this is what Saturn, Jupiter, Mars do. But
the others, Venus and Mercury, are moved around the sun, sometimes overtaking the sun, at other times [39] being overtaken by the sun, while never at-
taining an elongation of 60º from it.

This last clause was omitted by McColley from his quotation [40]
of Valla's Latin, just as he omitted Proclus-Valla's explanation of
the outer planets' elongations from the sun, ranging from 0º to 180º.
Hence McColley misunderstood Valla's *comprehendentes ... comprehensi*,
corresponding to Proclus' καταλαμβάνοντας ... καταλαμβανομένους [41]
(overtaking ... being overtaken). McColley's mistranslation (« Mer-

---

[36] McColley, p. 314/18-19.

[37] *Ibid.*, p. 313/1-2.

[38] *Ibid.*, p. 313/4.

[39] McColley mistranslated *modo ... modo* by « In one fashion ... in another way ».

[40] P. 314, n. 7.

[41] Ed. Manitius, p. 8/20-21.

cury and Venus encircle the sun ... they are encircled by the sun » [42])
confuses the pattern of Proclus-Valla's sun-Mercury-Venus with a
binary stellar system, whose components revolve around each other.
By contrast, Proclus-Valla's sun overtakes Mercury and Venus, and
is overtaken by them, so that they are sometimes morning stars and
at other times evening stars. They are not « encircled by the sun ».
Their orbits are centered on the sun, whereas the outer planets' orbits
are centered on the earth. The outer planets are geocentric, the accompanying planets are heliocentric.

By contrast, in Tycho Brahe's system, the revolutions of all five
planets were centered on the sun, whose orbit was centered on the
earth. Thus, his planetary system has been labeled ' geoheliocentric'. But in Proclus-Valla's geoheliocentric system, only the accompanying planets were heliocentric, the outer planets being geocentric. To apply the same undifferentiated label « geoheliocentric »
to two such different systems is to do the history of science the sort of
disservice of which McColley was a master.[43] In this instance, he said:

forces more powerful than the « inspiration » of Tycho Brahe made this
[geoheliocentric] theory a major hypothesis during the late sixteenth and
early seventeenth centuries ... [L]ong before Tycho the geo-heliocentric
theory had been the most active among the several competitors of conventional geocentric astronomy.[44]

The pre-Tychonic geoheliocentric theory, with its heliocentrism
restricted to Mercury and Venus, found only limited support. On
the other hand, a far more favorable response greeted Tycho's challenge to Copernicus' fully heliocentric astronomy. Its planetary moving earth displeased many, who welcomed Tycho's planetary heliocentrism because his earth continued to remain non-planetary, stationary, and central.

---

[42] McColley, p. 314/15-16.

[43] For other examples, see TCT, pp. 238-239, # 467-470.

[44] McColley, p. 314.

16

Was Copernicus' Revolutions Approved by the Pope?

The Renaissance may be regarded as the historical movement that initiated the trend of modern society toward secularism. As the humanist attitude spread, lay contributors to culture became more influential than they had been when clerics were dominant. Nevertheless, the cloistered clergy's participation in various humanist activities was far from negligible. The requisite assessment of "The Contribution of Religious Orders to Renaissance Thought and Learning" was recently begun by Paul Oskar Kristeller.[1] His selective list of humanistically productive monks and friars included the name of Giovanni Maria Tolosani, a member of the Dominican Order.

The exact date of Tolosani's birth is not known. Descended from a Tuscan family that lived and owned property near Florence, in 1487 as a teen-ager he entered the Dominican friary of St. Mark in that city, and then officially joined the Order on 28 June 1488. He spent more than sixty years as a Dominican, and when he was nearly an octogenarian, he died on 22 January 1549.[2]

At the time of the Fifth Lateran Council (1512-17), the pope appealed to the experts for suggestions about improving the calendar, which was known to be out of phase with the seasons. Many specialists replied, including Nicholas Copernicus, whose response has not been found.[3] By contrast Tolosani's essay *On the Correction of the Calendar (De correctione calendarii)* survives in Florence's National Central Library, which preserves the libraries of the local clerical Orders whose houses were shut down. During this process Tolosani's

[1]P. O. Kristeller, *Medieval Aspects of Renaissance Learning* (Durham, North Carolina, 1974), 93-158; originally published in *American Benedictine Review,* **21** (1970), 1-55.

[2]Demetrio Marzi, "La questione della riforma del calendario nel Quinto Concilio Lateranense (1512-1517)," *Pubblicazioni del r. Istituto di Studi Superiori in Firenze,* sezione di filosofia, **27** (1896), 130-131. The date of Tolosani's death was recorded in Serafino Loddi's manuscript entitled *Monumenta S. Marci de Florentia.* An eighteenth-century extract from Loddi's manuscript was used by Gustavo Galletti, when he added Tolosani's *La Nuova Sfera* to *La Sfera,* ascribed to Leonardo Dati (Florence, 1859; Milan, 1865). Loddi's manuscript was not used by Jacques Quétif and J. Echard, *Scriptores ordinis praedicatorum* (Paris, 1719-21; reprint, New York, 1959; II, 123), whose date (1545) for Tolosani's death is no better than a guess. On the other hand, "about 1595" in Kristeller's *Medieval Aspects,* 155, is admittedly a mere slip; see note 6, below.

[3]Eugenio Garin committed a twofold error in saying that "Copernicus declined the invitation of Paul of Middelburg, bishop of Fossombrone." First, the invitation was issued by the pope, not by Paul. Secondly, Paul's *Secundum compendium correctionis calendarii* (Rome, 1516, sig. b1r) listed Copernicus among those who responded to the papal invitation by writing to Rome rather than by appearing in Rome in person. See Garin, "A Proposito di Copernico," *Rivista Critica di Storia della Filosofia,* **26** (1971), 87; and Garin, "Alle Origini della Polemica Anticopernicana," *Studia Copernicana,* VI (Wrocław, 1973), 31.

essay was transferred from St. Mark to the National Central Library's *Fondo Conventi Soppressi,* where it reposes today.[4]

After the Fifth Lateran Council ended in 1517 without reaching a definite decision about reforming the calendar, further researches were encouraged by the papacy. In this context a collection of five chronological tracts was issued in September 1537 by the famous Venetian publisher Luca Antonio Giunti under the title *Opusculum de emendationibus temporum,* attributed to "Joannes Lucidus Samotheus."[5] This supposed author was referred to as a deceased Frenchman, in the dedication which was signed by Tolosani. But no trace of any such Frenchman, dead or alive, has ever been found.

Why did Tolosani conceal his authorship of the *Opusculum?* Inwardly he was intensely critical of *Paulina, de recta Paschae celebratione* (Fossombrone, 1513) by Paul of Middelburg (1445–1533). That formidable polemicist "was then in charge of this matter," as Copernicus said in dedicating to the pope his *De revolutionibus orbium coelestium* (more briefly, *Revolutions;* Nuremberg, 1543). Since overt objections to *Paulina* by a Dominican might involve the Order in regrettable controversy, Tolosani resorted to a pseudonym.[6] By the same token, in the following century, the Jesuit Christopher Scheiner, who became embroiled in a bitter dispute with Galileo, hid behind the pseudonym Apelles to avoid bringing discredit on his Order.

How did Tolosani choose the names of his pseudonym? In his dedication of the *Opusculum* Tolosani said that the pretended author "so elucidated" his subject "that he deservedly ought to be named Lucidus" (*ita elucidavit, ut merito Lucidus cognominari debeat*). A man's accomplishments during his lifetime may justify the addition of an epithet to the names he received at birth (e.g., Alexander the Great, Quintus Fabius Maximus the Delayer). But had there really been any such family name as Lucidus, should not the fictitious writer have been praised for living up to his inherited surname rather than for deserving the name Lucidus?

To "Joannes Lucidus," Tolosani's original christening appended "Censor Samotheus."[7] A censorious author, however, was more likely to promote discord than to reconcile conflicting opinions in the calendar debate. After further reflection, therefore, Tolosani prudently deleted "Censor" from the label of his phantom.

As for Samotheus, in the *Opusculum* (fol. 44r) Tolosani said: "The Celts or Gauls, the progenitors of the French, owed their origin to Samotes . . . after whom the Gauls were called Samotei." This much Tolosani quoted from a source whom he cited as "Berosus," and then added the gloss "Samotei, that is, wise." He based this gloss on his source's statement that the wise men "among the Celts and Gauls were the Druids and the so-called Samothei, who

[4] P. O. Kristeller, *Iter italicum* (London, 1963–67), I, 164: composite manuscript J VIII 9, fol. 256–59. Tolosani's essay was published by Marzi, "Questione," 250–54.

[5] Copy in New York Public Library, class mark *KB 1537.

[6] Demetrio Marzi, "Giovanni Maria Tolosani e Giovanni Lucido Samoteo," *Miscellanea storica della Valdelsa,* 5 (1897), 42–45, 50, 59. At p. 52 Marzi quotes Roberto Ubaldini's manuscript *Annalia conventus S. Marci de Florentia* for 22 January 1549 as the date, Florentine style, of Tolosani's death.

[7] Marzi, "Tolosani e Lucido," 44.

. . . were profoundly learned in divine and human law. And for this reason they were deeply devoted to religion and therefore they were called Samotei."

The foregoing astonishing gibberish commended itself to Tolosani because it was promulgated by a fellow-Dominican, Giovanni Nanni or Annius of Viterbo (1432-1502). This perpetrator of one of the most shameless forgeries of his age (*De commentariis antiquitatum,* Rome, 1498) professed to be culling from and commenting on various ancient writers, including Berosus. Nanni's weird fabrications about Samotheus (sig. R1v–2r) utterly deceived poor, trusting Tolosani. He fondly imagined that by adding Samotheus to "Joannes Lucidus (Censor)," he was turning his pseudonym into a wise Frenchman. That Frenchman, moreover, had better be dead. For if he were alive, he might be invited through Giunti or Tolosani to participate personally in a discussion of calendar reform. In that case "Samotheus' " refusal to attend would either affront a powerful personage or disclose Tolosani's secret.

When the second edition of "Lucidus' " *Opusculum* was issued by L. A. Giunti in Venice, the colophon was dated December 1545, but the title page said 1546.[8] In this 1546 edition of the *Opusculum,* to the five tracts of the 1537 edition there were added two pieces attributed without any subterfuge to Tolosani himself (fol. 198–210). The second of these subjoined pieces was entitled *Brevis annotatio emendatoria calendarii romani.* Here Tolosani tells us that the Vicar General of the Dominican Order on his way to the Council of Trent rested for several days in the friary of St. Mark in Florence. While there he ordered Tolosani, "already burdened by old age," to compose a brief note correcting the Roman calendar (fol. 199v). In compliance Tolosani finished the *Brevis annotatio* on 8 May 1545 and handed it to his Vicar General for submission to the Council of Trent.

Nearly four years later Tolosani died. Hence the third edition of his *Opusculum* (Venice, 1575) underwent no further major changes. But it appeared during the lively discussions that culminated in the Gregorian reform of 1582, which gave us our present calendar. This is based in large part on Tolosani.[9]

In Florence's National Central Library, Fondo Conventi Soppressi, manuscript J I 25 contains Tolosani's treatise *On the Truth of Holy Scripture.*[10] To this extensive work, which was completed in June 1544, Tolosani later added several appendices, of which the fourth, at fol. 339–43, may for brevity's sake be called *Heaven and the Elements.*[11] This fourth appendix provides hitherto unnoticed information about the Vatican's attitude toward Copernicus' *Revolutions,* the masterpiece which inaugurated the era of modern astronomy, if not of modern science as a whole. Tolosani's significant testimony was first brought to light by that illustrious and indefatigable scholar Eugenio Garin, who then proceeded to publish *Heaven and the Ele-*

[8]Copy in New York Public Library, class mark *KB 1546.

[9]Demetrio Marzi, "Giovanni Maria Tolosani, Alessandro Piccolomini e Luigi Giglio," *Miscellanea storica della Valdelsa,* 5 (1897), 202, 204, 208.

[10]Kristeller, *Iter,* I, 153: *De veritate sacrae scripturae.*

[11]Tolosani's short title: *De coelo et elementis;* his full title: *De coelo supremo immobili et terra infima stabili, ceterisque coelis et elementis intermediis mobilibus.*

ments in its entirety from Tolosani's own autograph.[12] In so doing Garin performed a notable service. The civilized world owes him a profound debt of gratitude for having made it possible at last to rectify a grave historical error of long standing.

This mistake occurs in the earliest surviving substantial biography of Copernicus, which was completed on 7 October 1588. Its author was the famous polymath Bernardino Baldi (1553–1617). Like any other reader of the front matter in the *Revolutions* as printed, Baldi had before him the letter sent to Copernicus by Nicholas Schönberg, cardinal of Capua, on 1 November 1536.[13] In the absence of any additional information Baldi wrote:

Schönberg had Copernicus' work; recognized its perfection and excellence; showed it to the pope, by whose judgment it was approved. The said Cardinal [Schönberg] addressed himself to Copernicus to ask him for many reasons to be willing to publish it.[14]

What induced Baldi to say that "Schönberg had Copernicus' work" before the cardinal wrote to the astronomer on 1 November 1536? Baldi surely knew that Cardinal Schönberg died on 9 September 1537. This interval of a little more than ten months between the Schönberg-Copernicus letter and the cardinal's death was hardly long enough to encompass travel from Rome to Poland and back again, with the copying of the *Revolutions* sandwiched in between the two trips. Hence Baldi dreamed up a previous Schönberg-Copernicus letter, which allegedly resulted in "Schönberg's having Copernicus' work." But in the only actual Schönberg-Copernicus letter, that dated 1 November 1536, the cardinal plainly says that he has heard about Copernicus' "new cosmology" and has

also learned that you have written an exposition of this whole system of astronomy. . . . Therefore with the utmost earnestness I entreat you . . . at the earliest possible moment to send me your writings on the sphere of the universe together with the tables and whatever else you have that is relevant to this subject. Moreover I have given instructions . . . to have everything copied in your quarters at my expense and dispatched to me.

Clearly, Schönberg did not have the *Revolutions* before he wrote to Copernicus on 1 November 1536.

Accordingly the rest of Baldi's contrived scenario disappears. Since Copernicus' *Revolutions* was not in the hands of Schönberg, Baldi's allegation that the cardinal "showed it to the pope, by whose judgment it was approved" vanishes in thin air. After saying what everybody knew about the *Revolutions,* that "Copernicus dedicated it to Pope Paul III," Baldi repeated his unfounded fabrication that "by the pope's judgment, as has been said, it had been approved."[15] Baldi then hastened to add: "What reward Copernicus obtained for

[12]These two valuable articles by Garin are cited above, note 3.

[13]About a year earlier, on 15 October 1535, Tolosani dedicated his *Opusculum* to this same Nicholas (Schönberg), cardinal of Capua.

[14]Bronislaw Biliński, *La Vita di Copernico di B. Baldi* (Wrocław, 1973), 22–23.

[15]Biliński, 23. In his biography of Paul of Middelburg, Baldi said: "The writings of that man [Lucido, whose identity with Tolosani was not understood by Baldi] did not fall into his [Paul's] hands, since, had he been aware of them, he would not have been able to refrain from defending himself in some way" (Marzi, "Questione," 245).

it and what happened in the said affair, I would not know." How much better the cause of historical accuracy would have been served had Baldi included the entire Schönberg-Copernicus-Paul III triangle within his forthright admission of ignorance!

Was it possibly through lack of conviction that Baldi avoided repeating his fictional tale about Pope Paul III's approval of Copernicus' *Revolutions* when the biographer had a suitable opportunity to do so? Baldi's first biography of Copernicus, completed in 1588, formed part of his *Vite de' matematici,* a manuscript too bulky to be published. Hence he produced compressed versions (*Cronica de' matematici,* Urbino, 1707). There, in his briefer biography of Copernicus, Baldi restricted himself to the readily verifiable statement that "Copernicus dedicated his great work on the *Revolutions* to Paul III" (*Cronica,* 121). But in the *Cronica* Baldi said absolutely nothing about any supposed approval of Copernicus' *Revolutions* by Pope Paul III.

Did Baldi perhaps later realize that he had previously confused Paul III with his immediate predecessor, Clement VII? The latter reacted quite favorably to an "explanation of Copernicus' opinion about the earth's motion." Copernicus' geokineticism was explained in the summer of 1533 in the Vatican gardens to Pope Clement VII and four of his associates. On that occasion Clement VII generously rewarded the pro-Copernican lecturer with the gift of a rare Greek manuscript. But Clement VII was a very different pope from Paul III. There is no evidence that Paul III ever listened to a pro–Copernican lecture or that Schönberg showed him Copernicus' *Revolutions* before it was first published in 1543. Baldi's unsupported assertion in the *Vite* that Paul III approved Copernicus' *Revolutions,* significantly enough, was not repeated by Baldi in his *Cronica.*

The damage done by Baldi's misstatement in the *Vite* cannot be accurately assessed because nobody knows with certainty who had access to that unpublished manuscript. At any rate it did not influence Galileo Galilei (1564–1642), whose pronouncement about our subject was quite different from Baldi's. For in his *Letter to the Grand Duchess,* while striving to tie Copernicus as tightly as possible to the Roman Catholic Church, Galileo (mistakenly) asserted that Copernicus undertook the heavy burden of writing the *Revolutions* "by order of the Supreme Pontiff."[16] Of course, no such order was ever given or received.

Nor is there any truth in the Baldi-like statement by Galileo's defender, Tommaso Campanella (1568–1639). In his *Apologia pro Galileo* (Frankfurt/Main, 1622), written while he languished in a Neapolitan jail, that courageous but uninformed Dominican said that "Pope Paul III ... approved" Copernicus' *Revolutions* and gave his "permission that the book should be printed" (pp. 9, 54). Neither in jail nor out of it could Campanella have found any evidence that Paul III approved or permitted the *Revolutions.*

If there were an iota of truth in the declarations by this celebrated trio of Baldi, Galileo, and Campanella that a pope had ordered, approved, or permitted the *Revolutions,* would not Copernicus have had that papal order, approval, or permission displayed even more prominently than Cardinal

---

[16] Edward Rosen, "Galileo's Misstatements about Copernicus," below, 196.

Schönberg's letter in the front matter of the *Revolutions?* Would not Copernicus, in his eloquent Dedication-Preface addressed to Pope Paul III, have expressed his deep gratitude for that papal order, approval, or permission, had any such ever been forthcoming?

These pronouncements by Baldi, Galileo, and Campanella were issued far from Rome some four to eight decades after the (pretended) event. Tolosani (c. 1471–1549), on the other hand, was almost exactly contemporary with Copernicus (1473–1543). Moreover, Tolosani was closely linked to the Vatican through Bartolomeo Spina of Pisa, who was appointed Master of the Sacred and Apostolic Palace by Pope Paul III in July 1542.

Because of Spina's outstandingly gifted mind, the pope thought most highly of him. Accordingly he relied on his advice in the difficult matters of faith which arose at that time, and the pope wanted Spina to be one of the five selected men whom the pope constituted at Rome to judge the questions raised at the Council of Trent.[17]

In his *History of the Council of Trent* Cardinal Sforza Pallavicino (1607–67) reports that on the Roman side

The leaders were mainly five in number: Francesco Romeo, general of the Dominicans, who used for this purpose two other theologians of his Order: Bartolomeo Spina, Master of the Sacred Palace. . . .[18]

When Tolosani was preparing to publish his treatise *On the Very Pure Truth of Divine Scripture, against Human Errors,*[19] he dedicated it to Pope Paul III, who ordered Spina to examine it. Spina's approval took the form of a letter to Tolosani, dated 16 August 1546. Therein Spina praised the work highly, commended it completely, and lauded its elegance, charm, and simplicity of exposition. Spina described himself as a fellow-Dominican and friend of Tolosani for fifty years.[20] This letter must have been written shortly before Spina died, since "he gave up his pious soul in 1546, being about 70 or 72 years old."[21]

We have now caught a glimpse of the lifelong friendship between Tolosani and Spina; of Pope Paul III's high regard for Spina's theological acumen; and of Spina's admiration for Tolosani. Let us next look more closely at Tolosani's aforementioned *Heaven and the Elements,* Chapter 2, where he refers to the *Revolutions,*

the book by Nicholas Copernicus of Toruń, which was printed not long ago and published in recent days. In it he tries to revive the teaching of certain Pythagoreans concerning the earth's motion, a teaching which had died out in

---

[17]Quétif-Echard, II, 126.

[18]S. Pallavicino, *Storia del Concilio di Trento,* VIII, 1 (Classici UTET, 2nd ed., Turin, 1968), 337–38.

[19]Florence, National Central Library, Corporazioni Religiose, S. Marco, J I 25, *De purissima veritate divine scripture adversus errores humanos,* fol. 3: Hoc egregium opus ab auctore manu sua exaratum, atque a Reverendissimo Magistro Sacri Palatii fra' Bartholomeo Pisano, Paulo 3° Pontifice mandante, approvatum. . . . This information was kindly obtained for me by Professor Vasco Ronchi, Director of the National Institute of Optics, Arcetri-Florence.

[20]Marzi, "Tolosani e Lucido," 51.   [21]Quétif-Echard, II, 126.

times long past. Nobody accepts it now except Copernicus. In my judgment, he does not regard that belief to be true. On the contrary, in this book of his he wanted to show others the keenness of his mind rather than expound the truth of the matter.

Since Tolosani, other anti-Copernicans have similarly made the baseless charge that Copernicus did not believe what he wrote. Tolosani continues:

As far as I could judge by reading his book, he is a man with a keen mind. He understands Latin and Greek, and expresses himself eloquently in those languages, not however without an obscurity in his phraseology since he uses unfamiliar words too often. He is also an expert in mathematics and astronomy, but he is very deficient in physics and dialectics. Moreover he seems to be unfamiliar with Holy Scripture since he contradicts some of its principles, not without the risk to himself and to the readers of his book of straying from the faith. . . .

Thoroughly familiar with the Bible, Copernicus farsightedly warned against allowing distortions of Biblical passages to impede the development of science. Tolosani continues:

Hence, since Copernicus does not understand physics and dialectics, it is not surprising if he is mistaken in this opinion and accepts the false as true, through ignorance of those sciences. Summon men educated in all the sciences, and let them read Copernicus, Book I, on the moving earth and the motionless starry heaven. Surely they will find that his arguments have no solidity and can be very easily refuted. For it is stupid to contradict a belief accepted by everyone over a very long time for extremely strong reasons, unless the naysayer uses more powerful and incontrovertible proofs, and completely rebuts the opposed reasoning. Copernicus does not do this at all. For he does not undermine the proofs, establishing necessary conclusions, advanced by Aristotle the philosopher and Ptolemy the astronomer.

Then let experts read Aristotle, *On the Heavens,* Book II, and the commentaries of those who have written about it . . . and they will find that Aristotle absolutely destroyed the arguments of the Pythagoreans. Yet this is not adduced by Copernicus in his ignorance of it, nor does he follow the Pythagoreans in all respects, since they put Fire in the middle near the center of the universe, where everybody else correctly and most convincingly proves that the earth is. Copernicus, however, puts the sun there, not Fire, and both are caught in a great error. For Copernicus puts the indestructible sun in a place subject to destruction. And since Fire naturally tends upward, it cannot, except through constraint, remain down near the center as its natural place, as the Pythagoreans falsely hold.

This difference between Copernicus and the Pythagoreans was ignored by many of Tolosani's contemporaries and coreligionists. The Pythagoreans' sun was a planet moving around the center of the universe, whereas Copernicus' sun was not a planet, did not move, and remained fixed at (or near) the center of the universe. At first, in his *Commentariolus* Copernicus conspicuously dissociated himself from the Pythagoreans. Later on, in the *Revolutions* he deemed it expedient to quote or cite ancient references to the Pythagoreans. These references induced less attentive students than Tolosani to lump Copernicus indiscriminately with the Pythagoreans. Tolosani himself, however, commits the same error later on, as we shall soon see. Meanwhile he continues:

Hence Copernicus, copying the Pythagoreans in part, leans on a cane of fragile reed which easily pierces his hand, or on an imaginary fabrication by which the truth cannot be proved. Therefore he is often mistaken. For in his imagination he changes the order of God's creatures in his system when, like the giant trying to pile Ossa on Pelion, he [seeks] to raise the earth, heavier than the other elements, from its lower place to the sphere where everybody by common consent correctly locates the sun's sphere, and to cast that sphere of the sun down to the place of the earth, contravening the rational order and Holy Writ, which declares that heaven is up, while the earth is down. . . .

Moreover Copernicus assumes certain hypotheses which he does not prove . . . when he says in Book I, Chapter 8: "If anyone believes that the earth rotates, surely he will hold that its motion is natural, not violent." Copernicus assumes what he should previously have proved, namely, that the earth rotates. This proposition, however, is explicitly shown to be false. For as far as a rotating earth is concerned, its motion cannot be called natural, but [must be called] coerced, since a simple body cannot have two natural motions opposed to each other. For we see the earth move naturally toward the center [of the universe] on account of its natural heaviness. But if it is said to rotate, its circular motion will be coerced, not natural. Therefore, it is false that the earth rotates with a natural motion. On the contrary, that motion is coerced, and thus Copernicus' hypothesis is completely overthrown.

Furthermore, in Book I, Chapter 10, this author falsely supposes that the "first and the highest of all [the spheres] is the sphere of the fixed stars, which contains itself and everything, and is therefore immovable." This is shown to be false, since the sphere of the fixed stars has two opposite motions, one natural, the other coerced. This could not be the case unless above it is the First Movable, which moves with a single, simple, uniform motion, as all informed astronomers agree. By the action of the First Movable, the starry heaven is moved contrary to its natural and proper motion. Copernicus would have spoken correctly, had he agreed with the theologians that above the First Movable the highest sphere is immovable, the sphere called by the theologians the Empyrean Heaven. This contains, as in an immovable place containing itself, all the lower movable heavenly spheres which revolve around the center of the universe. . . .

Almost all the hypotheses of this author Copernicus contain something false, and very many absurdities follow from them. Hence that writer, whose name is not indicated there, and who speaks "To the Reader Concerning the Hypotheses of This Work," although in the earlier part he flatters Copernicus, nevertheless toward the end of his remarks, viewing the truth of the matter correctly and without any adulation he says: "As far as hypotheses are concerned, let no one expect anything certain from astronomy, which cannot furnish it, lest he accept as the truth ideas conceived for another purpose, and depart from this study a bigger fool than when he entered it." This is what that unknown author says. These words of that author censure the book's lack of sense. For by a foolish effort it tries to revive the contrived Pythagorean belief, long since deservedly buried, since it explicitly contradicts human reason and opposes Holy Writ. Pythagoreanism could easily give rise to quarrels between Catholic expounders of Holy Writ and those persons who might wish to adhere with stubborn mind to this false belief. I have written this little work for the purpose of avoiding this scandal.

At this point Tolosani originally intended to terminate his *Heaven and the Elements*. Ignoring his previous distinction between Copernicanism and Pythagoreanism, Tolosani slips into that erroneous identification of the two which soon became the vogue, to the detriment of clear thinking. Even more noteworthy is Tolosani's recognition that the address "To the Reader Concerning the Hypotheses of This Work" was written by someone other than Copernicus. Had Tolosani's *Heaven and the Elements* been published by some alert predecessor of Garin, the history of ideas would have been spared many a sad spectacle due to the mistaken notion that the address "To the Reader" emanated from Copernicus himself. On the other hand Garin repeats a blunder concerning Osiander, the author of the address "To the Reader," and Copernicus' disciple Rheticus: "Rheticus . . . obtained from Osiander a written statement that the address 'To the Reader' was not by Copernicus, and although Rheticus did not publish Osiander's statement, he let it be known."[22] No such statement was ever written by Osiander or obtained by Rheticus.[23] After Chapter 2, with which Tolosani originally intended to conclude his *Heaven and the Elements,* he began Chapter 3 as follows:

Although I have ended my remarks, nevertheless, having been urged on by the advice of learned men, I think that some statements must still be added. For I have sent my reader to peruse the text of Aristotle, *On the Heavens,* Book II. It is not easy, however, for everybody to have that book in his own possession, together with the commentaries on it. Therefore, in order that readers may more readily learn that Nicholas Copernicus neither read nor understood the arguments of Aristotle the philosopher and Ptolemy the astronomer, for that reason I shall briefly adduce here their arguments and refutations of the opinion opposed by them. . . .

It is highly regrettable that Tolosani chose to conceal the names of the learned men who advised him to add to his first two Chapters. The accusation that Copernicus was unfamiliar with Aristotle is ludicrous.[24] Other anti-Copernicans have complained that Copernicus followed Ptolemy too closely. In his basic outlook Copernicus departed from Aristotle only where compelled to do so by his own geokineticism. Yet it has recently become fashionable to label Copernicus a Neoplatonist, Neopythagorean, or Hermetist. These groupings lean on a broken reed sliding on slippery mud, while ignoring Copernicus' firm attachment to the solid core of Aristotelianism.[25] Toward the end of Chapter 4 Tolosani remarks:

Read Book I of Nicholas Copernicus' *Revolutions,* and from what I have written here you will clearly recognize into how many and how great errors he has tumbled, even contrary to Holy Writ. Where he wished to show off the keenness of his mind in the book he published, by his own words and writings he rather revealed his own ignorance. Hence he has no right to complain about

[22]Garin, "Copernico e i filosofi italiani del Rinascimento," *Belfagor,* **28** (1973), 669.

[23]*Journal for the History of Astronomy,* 5 (1974), 202.

[24]Aleksander Birkenmajer, "Copernic philosophe," *Studia Copernicana,* IV (Wrocław, 1972), 614–24.

[25]Edward Rosen, "Was Copernicus a Hermetist?" *Minnesota Studies in the Philosophy of Science,* V (Minneapolis, 1970), 164–71.

the men with whom he disputed at Rome, and by whom he was most severely condemned. On the contrary it is more appropriate for him to thank those from whom he learned what he did not know. But that discussion took place belatedly or after the printing and publication of his book. And therefore the falsehoods he wrote had to be refuted by the truth in this little tract of mine, lest the readers of his book should be led astray by his aforementioned errors.

According to Tolosani's garbled account Copernicus disputed in Rome "belatedly" (*tarde*, too late to affect the contents of the *Revolutions*) "or after the printing and publication of his book." But Copernicus was in Rome only once in his life, more than four decades earlier, at the turn of the century. "After the printing and publication of his book" about 15 March 1543, his last illness confined him to his bed in Frombork, far from Rome, until his death on 24 May 1543.

Thereafter a close friend of his was in Rome (as was pointed out to me privately by Paul Lawrence Rose). Alexander Scultetus was respected by Copernicus as "the only one outstanding in every respect among all the officials and canons of our" Cathedral Chapter of Varmia. By the same token "Nicholas Copernicus, canon of Frombork, astronomer and mathematician" was listed in Scultetus' *Chronology or Annals of Nearly All the Kings, Princes and Potentates from the Creation of the World to the Year 1545* (Rome, 1546). But a heretical

book read and annotated by Alexander [Scultetus] was turned over to the principal theologian, who in Rome is called the Master of the Sacred Palace, so that he might read it and render a judgment as to what should be decided about it and the annotations and the individual who used this book.[26]

This judgment was rendered by the "Pisan Bartolomeo [Spina, Friar] of the Dominican Order, Master of the Sacred Palace" on 17 October 1545.[27] Spina must have reported this affair to Tolosani, and one of these two oldsters transformed Scultetus, who did dispute at Rome with men who condemned him most severely, into Copernicus. Tolosani concluded his *Heaven and the Elements* as follows:

The Master of the Sacred and Apostolic Palace had planned to condemn his [Copernicus'] book. But, prevented at first by illness, then by death, he could not carry out this [plan]. This I took care to accomplish afterwards in this little work for the purpose of safeguarding the truth to the general advantage of Holy Church.

Tolosani, privy to the Vatican's intentions, informs us that the Sacred Palace was planning to condemn Copernicus' *Revolutions* shortly after its publication, just as Galileo's *Dialogue* was later "prohibited by public edict." Whereas Galileo was sentenced to prison, Copernicus went scot-free by delaying the publication of his *Revolutions* until he was safely beyond the reach of the ecclesiastical authorities. The *Revolutions* itself was saved from condemnation by, first, the illness and, then, the death of the Master of the Sacred

---

[26] *Acta historica res gestas Poloniae illustrantia,* IV (Cracow, 1879) = *Stanislai Hosii . . . epistolae,* I, 200.   [27] *Ibid.,* I, 425–26.

Palace. Thereafter the Vatican was so deeply engrossed in the portentous Council of Trent that Copernicus' *Revolutions* escaped official notice.

An invidious comparison has often been made between the Protestant leaders Luther and Melanchthon, who spoke out against Copernicus, and the Catholic Church, which kept quiet. This silence has even been interpreted as benign approval. Outspoken approval, as we saw above, has also been alleged. Now, thanks to Garin's assiduity, Tolosani's testimony is available and shows that what happened later to Galileo's *Dialogue* nearly happened to Copernicus' *Revolutions.*

Tolosani's aforementioned manuscript *On the Very Pure Truth* carries an annotation, the beginning of which is quoted above in note 19. Where that quotation is interrupted, the annotation continues with the information that this Tolosani manuscript "was read aloud in public by Friar Tommaso Caccini, theologian of the famous Florentine Academy, in 1635."[28] Two decades earlier this same Friar Caccini, a Dominican of St. Mark in Florence, had gone to Rome in order to denounce Galileo's Copernicanism to the Inquisition.[29] Although the secret testimony he gave on Friday, 20 March 1615, was later impugned,[30] nevertheless he was largely responsible for the proceedings that culminated in Galileo's imprisonment and compulsory public renunciation of Copernicanism. By thus contributing to the condemnation of Galileo, Caccini helped to fulfill what had been left undone by Spina, whose friend Tolosani is an unimpeachable witness that Copernicus' *Revolutions* was not approved by the pope.[31]

[28] "1655" in Marzi, "Tolosani e Lucido," 41, is an error. Caccini died on 12 January 1648 (Quétif-Echard, II, 559).

[29] Antonio Favaro, *Galileo e l'inquisizione* (Florence, 1907), 15, 47–51.

[30] Antonio Ricci-Riccardi, *Galileo Galilei e Fra Tommaso Caccini* (Florence, 1902), 128.

[31] Besides Luther and Melanchthon, a third prominent contemporary Protestant leader used to be named as having emphatically disapproved of Copernicus. "Who will venture to place the authority of Copernicus above that of the Holy Spirit?" was a challenge formerly ascribed to John Calvin. But the readers of this *Journal,* 21 (1960), 431–41; 22 (1961), 386–88, were told that this rhetorical question was misattributed to Calvin, whose voluminous writings do not provide any substantial indication that he ever heard of Copernicus. Recently, however, attention was directed to a sermon preached by Calvin, probably in 1556, thirteen years after the publication of Copernicus' *Revolutions.* In that sermon Calvin denounced those who "pervert the order of nature. We shall see some of them so crazy . . . that they will say that the sun does not move and that it is the earth which revolves and that it turns" (*Revue de l'Histoire des Religions,* 179 [1971], 38). This passage has been taken to be "incontrovertible proof that Calvin did know the Copernican system and that he was dead against it" (Reijer Hooykaas, "Calvin and Copernicus," *Organon,* 10 [1974], 140). But the Copernican system was not exactly the same as the traditional geokinetic cosmology, which Calvin knew well. On the one hand, Copernicus agreed with his ancient predecessors in depriving the sun of motion, and in bestowing upon the earth both an annual revolution around the stationary sun and a daily rotation about its own axis. To these two familiar motions of the earth, however, Copernicus added a third, which was

absolutely novel, namely, the movement of the earth's axis. This axial motion of the earth was unknown to Copernicus' forerunners and also to Calvin. His perverters of the order of nature included the pre-Copernican geokineticists with their twofold motion of the earth, but not Copernicus himself, whose *Revolutions,* Book I, Chapter 11, is entitled "Proof of the Earth's Triple Motion" (Edward Rosen, "Calvin n'a pas lu Copernic," *Revue de l'Histoire des Religions,* **182** [1972], 183–84).

17

Calvin's Attitude towards Copernicus

> There will be a time when false and unprofitable
> professors will be made manifest and discovered.[1]

Bertrand Russell, in a book which has achieved an extensive distribution, quoted Luther's well-known denunciation of Copernicus as a fool, and then added:

Calvin, similarly, demolished Copernicus with the text: " The world also is stablished, that it cannot be moved " (Ps. 93:1), and exclaimed: " Who will venture to place the authority of Copernicus above that of the Holy Spirit? " [2]

Russell put this exclamation in Calvin's mouth without indicating on what occasion the Genevan reformer expressed his attitude toward the founder of modern astronomy.[3] This absence of any reference to a source for Calvin's exclamation should not surprise us, since Russell's *History of Western Philosophy* was " originally designed and partly delivered as lectures at the Barnes Foundation in Pennsylvania," [4] where the students' primary interest is focused on a topic somewhat removed from the development of speculative thought. Moreover, in his preface Russell acknowledged that he owed

a word of explanation and apology to specialists on any part of my enormous subject. It is obviously impossible to know as much about every philosopher as can be known about him by a man whose field is less wide; I have no doubt that every single philosopher whom I have mentioned, with the exception of Leibniz, is better known to many men than to me. If, however, this were considered a sufficient reason for respectful silence, it would follow that no man should undertake to treat of more than some narrow strip of history. . . . On such grounds I ask the indulgence of those readers who find my knowledge of this or that portion of my subject less adequate than it would have been if there had been no need to remember " time's winged chariot." [5]

To withhold our indulgence from Russell would surely be ungracious. Nevertheless it is difficult to repress the feeling of curiosity aroused by his tantalizing omission of documentary support for Calvin's anti-Copernican exclamation. I for one, as a student of early modern science, would be happy to learn exactly when and where Calvin exclaimed against Copernicus.

[1] John Owen, *An Exposition of the Epistle to the Hebrews* (2d ed., Edinburgh, 1812–1814), IV, 605; Observation VII on 5:12–14.

[2] Russell, *A History of Western Philosophy* (New York, 1945; London, 1946), 528; Italian tr. by Luca Pavolini, *Storia della filosofia occidentale* (Milan, 1948), 3 vols.

[3] This lack of documentation did not prevent an American Catholic philosopher from accepting Russell's remarks about Calvin at face value; see Pierre Conway, O.P., " Aristotle, Copernicus, Galileo," *New Scholasticism* XXIII (1949), 58.

[4] Russell, *Hist. West. Phil.*, xi.   [5] *Op. cit.*, x–xi.

This information was not furnished by Russell when, in an earlier work which has been re-issued five times, most recently in 1956, he ascribed the same anti-Copernican exclamation to Calvin.[6] Neither in his *Religion and Science* (1935) nor in his *History of Western Philosophy* (1945) did Russell evince any first-hand acquaintance with writings by Calvin, as distinguished from writings about Calvin. But more than once in his *Religion and Science* Russell quoted from A. D. White, *A History of the Warfare of Science with Theology in Christendom,* " to which," Russell admitted, " I am much indebted." [7]

Andrew Dickson White (1832–1918) was a historian who helped to found Cornell University and became its first president. When sectarian criticism of the new university intensified, White tells us,

having been invited to deliver a lecture in the great hall of the Cooper Institute at New York, [I] took as my subject " The Battle-fields of Science." In this my effort was to show how, in the supposed interest of religion, earnest and excellent men, for many ages and in many countries, had bitterly opposed various advances in science and in education, and that such opposition had resulted in most evil results, not only to science and education, but to religion. This lecture was published in full, next day, in the *New York Tribune;* extracts from it were widely copied; it was asked for by lecture associations in many parts of the country; grew first into two magazine articles, then into a little book which was widely circulated at home, reprinted in England with a preface by Tyndall, and circulated on the Continent in translations, was then expanded into a series of articles in the *Popular Science Monthly,* and finally wrought into my book on *The Warfare of Science with Theology.*[8]

In his " little book " (1876) White mentioned Calvin twice: once to recall that the Jesuits deemed Galileo's *Dialogue* " more pernicious for Holy Church than the writings of Luther and Calvin," [9] and once to praise Calvin for having drawn a distinction between lawful interest on a loan and oppressive usury.[10] Many years later, in the articles which White contributed to *Popular Science Monthly,* he introduced the following passage:

While Lutheranism was thus condemning the theory of the earth's movement, other branches of the Protestant Church did not remain behind. Calvin took the lead, in his *Commentary on Genesis,* by condemning all who asserted that the earth is not at the centre of the universe. " Who," he said, " will venture to place the authority of Copernicus above that of the Holy Spirit? " [11]

[6] Russell, *Religion and Science* (London, 1935), 23; (New York, 1935), 20.

[7] *Op. cit.* (London), 84; (New York), 86.

[8] *Autobiography of Andrew Dickson White* (New York, 1905), I, 425.

[9] *Le Opere di Galileo Galilei,* national ed. (Florence, 1890–1909; reprinted, 1929–1939), XV, 25; cf. XVI, 458.

[10] White, *The Warfare of Science* (New York, 1876; re-issued, New York, 1893), 36, 129.

[11] White, " New Chapters in the Warfare of Science," *Popular Science Monthly* XL (1891–1892), 587.

The latter part of this passage was then expanded by White in his *Warfare of Science with Theology* (1896) to read as follows:

Calvin took the lead, in his *Commentary on Genesis*, by condemning all who asserted that the earth is not at the centre of the universe. He clinched the matter by the usual reference to the 'first verse of the ninety-third Psalm, and asked, " Who will venture to place the authority of Copernicus above that of the Holy Spirit? " [12]

If we now compare this final version of White's comments about Calvin with our opening quotation from Russell, we shall feel fully justified in concluding that it was from White, not from Calvin, that Russell took the anti-Copernican exclamation which interests us.

Russell was of course not the only writer who borrowed from White. For instance, William Ralph Inge, Dean of St. Paul's in London, asserted that " Calvin asked, ' Who will venture to place the authority of Copernicus above that of Holy Scripture? ' " [13] Dean Inge's substitution of " Holy Scripture " for " the Holy Spirit " need not deter us from recognizing his dependence on White. The latter was quoted more faithfully by a director of the Lick Observatory,[14] by some American historians of science,[15] and also by a highly successful popularizer.[16] Lastly, in his recent volume on *The Reformation,* Will Durant rated Calvin " ahead of his time in doubting astrology, abreast of it in rejecting Copernicus." [17]

These nine diverse authors may serve to exemplify the strong, broad, and enduring influence exerted by White's *Warfare* which, since its first publication in 1896, has been frequently re-issued, most recently in 1955. But if we want to know the source of the anti-Copernican exclamation attributed to Calvin, we shall not find the answer to our question in White. He, unlike Russell, was a professional historian. In his effort to comprehend Calvin's mentality, White read not only books written about the Genevan reformer, but also treatises composed by Calvin himself, including his *Commentary*

[12] White, *A History of the Warfare of Science with Theology in Christendom* (New York, 1896), I, 127.

[13] *Science, Religion, and Reality,* ed. Joseph Needham (New York, 1925), 359.

[14] Edward Singleton Holden, " Copernicus," *Popular Science Monthly* LXV (1904), 120: " To Calvin the pronouncement of Copernicus was sheer blasphemy. It seemed to him to lie entirely within the sphere of religion. Judged by the accepted standards of that sphere it was audacious heresy " (reprinted in *Scientific American Supplement* LVIII [1904], 24068).

[15] Henry Smith Williams, *The Great Astronomers* (New York, 1930), 236; Benjamin Ginzburg, *The Adventure of Science* (New York, 1930), 95; Grant McColley, " The Ross-Wilkins Controversy," *Annals of Science* III (1938), 166, with the 1919 re-issue of White's *Warfare* cited on p. 154; Thomas S. Kuhn, *The Copernican Revolution* (Cambridge, Mass., 1957; reprinted, New York, 1959), 192, with White cited at p. 281.

[16] Hermann Kesten, *Copernicus and His World,* English tr. by E. B. Ashton and Norbert Guterman (New York, 1945; London, 1946), 315–316; French tr. by Eugène Bestaux, *Copernic et son temps* (Paris, 1951).

[17] Durant, *The Reformation* (New York, 1957; *The Story of Civilization,* VI), 477, with the 1929 re-issue of White's *Warfare* listed in the bibliography on p. 952.

on Genesis.[18]   Yet when White assigned the anti-Copernican exclamation to
Calvin, he did not specify as his source any of the reformer's writings.   Instead, he said: " On the teachings of Protestantism as regards the Copernican theory, see citations in Canon Farrar's *History of Interpretation*, preface, xviii." [19]

Frederic William Farrar (1831–1903), in his long and distinguished
church career, was at various times canon and archdeacon of Westminster,
dean of Canterbury, and chaplain of the House of Commons as well as of
Queen Victoria.   In recognition of his ecclesiastical attainments he was invited to deliver the Bampton Lectures in 1885.   More than a century before,
the Rev. John Bampton, canon of Salisbury, had bequeathed his estate to
Oxford University for the purpose of endowing annual lectures, the first of
which were delivered in 1780.   Choosing the development of Biblical exegesis as the subject of his Bampton Lectures in 1885, Farrar upheld the freedom of interpretation.   He zealously maintained the traditional Protestant
right of private judgment, aware of but not bound by received opinion.
Hence he delighted in cataloguing the errors of earlier theologians: Lactantius ridiculed the idea that the earth was round, Augustine denied that
there were antipodes, and " ' Who,' asks Calvin, ' will venture to place the
authority of Copernicus above that of the Holy Spirit? ' " [20]

This anti-Copernican exclamation was taken, as we observed above, by
Russell from White, and by White in turn from Farrar.   But here our trail
ends in a blind alley, Farrar having failed to say where he found the exclamation which he imputed to Calvin.

If we wish to understand the reason for this failure on Farrar's part, let
us listen to his son and biographer:

In judging Farrar's work, and this is true not only of the *Life of Christ*, but
of all his books, it must not be forgotten that there are two orders of
scholars, the " intensive " and the " extensive " school, both necessary to the
world—those whose function is original research, and those whose function
it is to interpret and make available the labors of the former class, whose
work would otherwise remain buried under its own weight. And it was to
this latter class that my father unquestionably belonged.[21]

Farrar's son further informs us that his father was " up to the eyes in pas-
toral and literary work, [and] burdened with a large correspondence." [22]
Not only was Farrar overworked, but

expression was easy to him; he poured out his ardent soul as the Spirit gave
him utterance, and without effort lavished from the rich treasures of his
memory garnered stores of poetic illustration and historic parallel. . . .
Quotation with him was entirely spontaneous, almost involuntary, because
his marvelous memory was stored, nay, saturated with passages.[23]

[18] *Warfare*, I, 10, 28.

[19] *Warfare*, I, 128; at I, 83–84, White lauded Farrar's behavior during the
controversy over Darwin's theory of evolution; for another eulogy of Farrar's courage, see White, II, 206.

[20] Farrar, *History of Interpretation* (London, 1886; New York, 1886), xviii.

[21] Reginald Farrar, *The Life of Frederic William Farrar* (New York, 1904), 193.

[22] *Op. cit.*, 255.   [23] *Op. cit.*, 255–256.

Even if we make every allowance for filial devotion, this portrait of a non-intensive scholar, overburdened with work, facile in expression, and relying on a marvelous memory for his quotations, inevitably creates a certain doubt in our minds. Did Farrar perhaps read in some other author the anti-Copernican exclamation to which he attached the name of Calvin? Our doubt grows deeper when we recall that in his preface Farrar candidly declared:

In a work which covers such vast periods of time and which involves so many hundreds of references it would be absurd to suppose that I have escaped from errors. All that I can say is that in this, as in my other works, I have done—not perhaps the best that I might have done under more favorable conditions of leisure and opportunity—but the best that was possible to me under such circumstances as I could command . . . . I beg the indulgent consideration of all who believe that I am actuated solely by the desire to do nothing against the truth.[24]

Despite this forthright disclaimer of infallibility, Farrar's unsupported ascription of the anti-Copernican exclamation to Calvin was repeated without hesitation not only by White and those who followed him, but also by Miss Dorothy Stimson, former dean of Goucher College. After quoting the exclamation from Farrar, Miss Stimson asserted that " Luther, Melanchthon, Calvin, Turrettin, Owen, and Wesley are some of the notable opponents " [25] of Copernicanism. Her assertion was echoed by the foremost German historian of astronomy,[26] from whom it passed into a biographical dictionary of German history.[27] The grave danger that the German-speaking world may now in the twentieth century begin to look upon Calvin as an outspoken adversary of the Copernican theory was increased by the publication of the German version of Kesten's popular book.[28]

As we have just seen, Miss Stimson accepted without reservation Farrar's attribution of the anti-Copernican exclamation to Calvin. But she balked at the following statement by Farrar: " ' Newton's discoveries,' said the Puritan John Owen, ' are against evident testimonies of Scripture.' " [29] Reporting that she was " unable to verify this statement," Miss Stimson astutely added that " Owen died [in 1683, four years] before the *Principia* was published in 1687." [30]

[24] F. W. Farrar, *History of Interpretation*, xxix.

[25] Dorothy Stimson, *The Gradual Acceptance of the Copernican Theory of the Universe* (Hanover, New Hampshire, 1917; New York, 1917), 41, 99. Although Miss Stimson listed (p. 131) the 1898 re-issue of White in her bibliography, and at pp. 100–104 cited White's discussions of other writers, for Calvin she relied on Farrar. She in turn was followed by the popular biographer Phillips Russell, *Harvesters* (New York, 1932), 75, with Miss Stimson cited at p. 301.

[26] Ernst Zinner, *Geschichte und Bibliographie der astronomischen Literatur in Deutschland zur Zeit der Renaissance* (Leipzig, 1941), 33; at p. 86 Zinner listed Miss Stimson's *Gradual Acceptance* among the works which he utilized.

[27] Hellmuth Rössler and Günther Franz, *Biographisches Wörterbuch zur deutschen Geschichte* (Munich, 1952), 478.

[28] Kesten, *Copernicus und seine Welt* (Amsterdam, 1948), 381; cf. n. 16, above.

[29] Farrar, *Hist. of Interpretation*, xviii.   [30] Stimson, *Gradual Acceptance*, 89.

Miss Stimson somehow failed to notice that Farrar treated Owen and Calvin differently. On the one hand, Farrar offered no source whatever for Calvin's anti-Copernican exclamation. On the other hand, according to Farrar, " John Owen (*Works*, XIX, 310) said that Newton's discoveries were ' built on fallible phenomena, and advanced by many arbitrary presumptions against evident testimonies of Scripture.' " [31] The words enclosed within single quotation marks were indeed used by Owen.[32] But his target was " the late hypothesis, fixing the sun as in the centre of the world," in other words, pre-Newtonian Copernicanism. Owen was not talking about " Newton's discoveries," despite Farrar's two misstatements to that effect.

This, then, is the sort of broken reed on which White leaned when he repeated Farrar's attribution of the anti-Copernican exclamation to Calvin. As we have already seen, Farrar did not designate any of Calvin's writings as the source of this exclamation. On the other hand, as the attentive reader will doubtless recall, one of Calvin's numerous publications was singled out by White when he declared that

while Lutheranism was thus condemning the theory of the earth's movement, other branches of the Protestant Church did not remain behind. Calvin took the lead, in his *Commentary on Genesis*, by condemning all who asserted that the earth is not at the centre of the universe.[33]

As was pointed out above, White was personally familiar with Calvin's *Commentary on Genesis*.[34] Yet it was not this work which White cited in his discussion of Protestant opposition to Copernicanism. On the contrary, he referred his readers to " Rev. Dr. Shields, of Princeton, *The Final Philosophy*, pp. 60, 61." [35]

Charles Woodruff Shields (1825–1904), " professor of the harmony of science and revealed religion in Princeton University," [36] published *The Final Philosophy* in the firm conviction that science and religion could be reconciled. Like his British counterpart Farrar, Shields found fault with earlier theologians who had impeded the advancement of science. To illustrate this regrettable tendency, Shields stated that

Calvin introduced his *Commentary on Genesis* by stigmatizing as utter reprobates those who would deny that the circuit of the heavens is finite and the earth placed like a little globe at the centre.[37]

[31] Farrar, *Hist. of Interpr.*, 432.

[32] *Exercitations concerning the Name, Original, Nature, Use, and Continuance of a Day of Sacred Rest*, Exercitation II = *An Exposition of the Epistle to the Hebrews*, Exercitation XXXVI, section 16 (*Works*, London, 1850–1855; re-issued, Edinburgh, 1862, XIX, 310).

[33] White, *Warfare*, I, 127.   [34] See n. 18, above.

[35] White, *Warfare*, I, 128; cf. I, 148; at I, 234 White cited Shields' " very just characterization of various schemes " for reconciling Genesis with geology.

[36] Shields so described himself on the title page of his book, *The Scientific Evidences of Revealed Religion* (New York, 1900).

[37] Shields, *The Final Philosophy* (New York, 1877; 2d ed., New York, 1879), 60; *Philosophia ultima* (New York, 1888–1905), I, 61.

Evidently White's assertion that "Calvin . . . in his *Commentary on Genesis* . . . condemn[ed] all who asserted that the earth is not at the centre of the universe" was based on Shields.

In trusting Shields, was White leaning on a stouter reed than Farrar proved to be? [38] With a view to answering this question, let us look at the following sentence in the Argument with which "Calvin introduced his *Commentary on Genesis*": "We indeed are not ignorant that the circuit of the heavens is finite, and that the earth, like a little globe, is placed in the centre." [39] "That the circuit of the heavens is finite and the earth placed like a little globe at the centre" is Shields' unmistakable echo. But Shields spoke of Calvin "stigmatizing as utter reprobates those who would deny that . . . the earth [is] placed like a little globe at the centre." In this passage of the Argument introducing his *Commentary on Genesis* Calvin did indeed speak harshly about persons professing certain other ideas, but the notion that the earth was at the centre of the universe was confidently affirmed by Calvin without referring to those who would deny it. Despite Shields, in this passage Calvin was not "stigmatizing as utter reprobates those who would deny that . . . the earth [is] placed like a little globe at the centre." As a reed on which to lean, Shields turns out to have been no less broken than Farrar.[40]

It was from Farrar that White took the unsupported anti-Copernican exclamation attributed to Calvin. It was from Shields that White took the misstatement about the condemnation of Copernicanism in the Argument introducing Calvin's *Commentary on Genesis*. But we should now recall

[38] "Owen . . . declared the Copernican system a 'delusive and arbitrary hypothesis, contrary to Scripture,'" said White (*Warfare*, I, 128), echoing Shields (*Final Philosophy*, 60–61; *Philosophia ultima*, I, 61–62). "Owen declared that Newton's discoveries were 'built on fallible phenomena and advanced by many arbitrary presumptions against evident testimonies of Scripture,'" said White (*Warfare*, I, 148), echoing Farrar (432). This anti-Newtonianism ascribed to Owen by Farrar and White was correctly discarded by Miss Stimson on chronological grounds. But was Owen as singlemindedly anti-Copernican as is implied by the curtailed quotation in Shields (echoed by White and by Henry Smith Williams, *Great Astronomers*, 235)? Owen goes on at once to say that, apart from theological considerations, there are against Copernicanism "reasons as probable as any that are produced in its confirmation . . . for it is certain that all the world in former ages was otherwise minded; and our argument is not taken, in this matter, from what really *was true*, but from what was universally apprehended *so to be.*" Do not Owen's italics indicate his awareness that "what really was true" was Copernicus' rearrangement of the planets around the sun?

[39] Calvin, *Commentaries on the First Book of Moses called Genesis*, tr. John King (Edinburgh, 1847–1850), I, 61; re-issued, Grand Rapids, 1948. Calvin's works were collected in *Corpus reformatorum*, XXIX–LXXXVII (Braunschweig and Berlin, 1863–1900) = *Ioannis Calvini opera quae supersunt omnia*, I–LIX; for this passage, see *Calvini opera*, XXIII, 10.

[40] Yet Shields' thorough knowledge of Calvin's life and teachings was amply demonstrated in two spirited defences of Calvin's character and conduct, "The Doctrine of Calvin Concerning Infant Salvation" and "The Trial of Servetus," *Presbyterian and Reformed Review*, I (1890), 634–651, IV (1893), 353–389, as well as in Shields' play, *The Reformer of Geneva* (New York and London, 1898).

that White included the following third element in his formulation of Calvin's attitude toward Copernicus: " He clinched the matter by the usual reference to the first verse of the ninety-third Psalm." [41] With deplorable reticence White declines to tell us where Calvin cited Ps. 93:1 against Copernicus.

In the absence of such guidance from White, let us look at Calvin's commentary on Ps. 93:1, the relevant portion of which reads as follows:

The heavens revolve daily, and, immense as is their fabric and inconceivable the rapidity of their revolutions, we experience no concussion—no disturbance in the harmony of their motion. The sun, though varying its course every diurnal revolution, returns annually to the same point. The planets, in all their wanderings, maintain their respective positions. How could the earth hang suspended in the air were it not upheld by God's hand? By what means could it maintain itself unmoved, while the heavens above are in constant rapid motion, did not its Divine Maker fix and establish it? [42]

We see that in Calvin's cosmology " the heavens revolve daily " [43] and the sun has a " diurnal revolution." These two phenomena were interpreted by Copernicus as mere appearances caused by the real rotation of the earth around its axis once every twenty-four hours. Furthermore, in Calvin's conception, the sun had an annual motion, which likewise was transformed by Copernicus into an optical illusion produced by the actual revolution of the earth around the sun once a year. The annual revolution and the daily rotation were assigned by Copernicus to the earth, whereas Calvin believed the earth to " hang suspended in the air " and to " maintain itself unmoved." Clearly, Calvin's cosmology was pre-Copernican. But was it anti-Copernican? Although we are told that Calvin " quoted Psalm 93 against Copernicus," [44] we look in vain for any anti-Copernican utterance in Calvin's commentary on Ps. 93.

Theological opponents of Copernicus usually quoted Ps. 104:5 (God " laid the foundations of the earth, that it should not be removed for ever "), on which Calvin's comment begins as follows:

Here the prophet celebrates the glory of God, as manifested in the stability of the earth. Since it is suspended in the midst of the air, and is supported only by pillars of water, how does it keep its place so stedfastly that it cannot be moved? This I indeed grant may be explained on natural principles; for the earth, as it occupies the lowest place, being the centre of the world, naturally settles down there. [45]

Here we are reminded that Calvin believed the earth to be " the centre of

[41] White, *Warfare*, I, 127.

[42] Calvin, *Commentary on the Book of Psalms*, tr. James Anderson (Edinburgh, 1845–1849), IV, 6–7; re-issued, Grand Rapids, 1949; *Calvini opera*, XXXII, 16–17.

[43] Commenting on Ps. 19:4, Calvin says: " the firmament, by its own revolution, draws with it all the fixed stars " (*Commentary on . . . Psalms*, I, 315; *Calvini opera*, XXXI, 198).

[44] James Gerald Crowther, *Six Great Scientists* (London, 1955), 36.

[45] Calvin, *Commentary on . . . Psalms*, IV, 148–149; *Calvini opera*, XXXII, 86.

the world," where Copernicus stationed the sun. For Calvin, the motionless earth in the centre of the world' posed a problem which he discussed as follows in his comment on Ps. 75:3: " The earth occupies the lowest place in the celestial sphere, and yet instead of having foundations on which it is supported, is it not rather suspended in the midst of the air? " [46] This concept of a stationary earth at rest in the air was picturesquely elaborated by Calvin in his sermon on the alphabetical Ps. 119:90 (" Thou hast established the earth, and it abideth "): [47]

I beseech you to tell me what the foundation of the earth is. It is founded both upon the water and also upon the air: behold its foundation. We cannot possibly build a house fifteen feet high on firm ground without having to lay a foundation. Behold the whole earth founded only in trembling, indeed poised above such bottomless depths that it might be turned upside down at any minute to become disordered. Hence there must be a wonderful power of God to keep it in the condition in which it is. [48]

The Aristotelians explained the steadiness of the earth in a naturalistic way that was familiar to Calvin who, however, found their explanation inadequate:

The philosophers indeed hold that the earth stands naturally in the middle of creation, as it is the heaviest element; and the reason they give that the earth is suspended in mid-air, is, because the centre of the world attracts what is most heavy; and these things indeed they wisely discuss. Yet we must go further; for the centre of the earth is not the main part of creation; it hence follows that the earth has been suspended in the air, because it has so pleased God. [49]

On another occasion Calvin invoked the idea of an orderly cosmos to account for the stability of the earth, and he also took advantage of this opportunity to paint a vivid picture of the universe as a huge sphere. Preaching on the text of Job 26:7 (" He stretcheth out the north over the empty place, and hangeth the earth upon nothing "), Calvin declared:

It is true that Job specifically says " the north," and yet he is speaking about the whole heaven. And that is because the sky turns around upon the pole that is there. For, just as in the wheels of a chariot there is an axle that runs through the middle of them, and the wheels turn around the axle by reason of the holes that are in the middle of them, even so is it in the skies. This is manifestly seen; that is to say, those who are well ac-

[46] Calvin, *Commentary on . . . Psalms*, III, 186; *Calvini opera*, XXXI, 702.

[47] In Calvin's *Commentary on . . . Psalms*, IV, 469 (*Calvini opera*, XXXII, 253), this verse evoked only the brief remark " that the earth continues stedfast, even as it was established by God at the beginning."

[48] Calvin, *Two and Twentie Sermons . . . [on] the Hundredth and Nineteenth Psalme*, tr. Thomas Stocker (London, 1580), fol. 99v (somewhat modernized); *Calvini opera*, XXXII, 620: sermon 12, preached on April 9, 1553.

[49] Calvin, *Commentaries on the Book of the Prophet Jeremiah and the Lamentations*, tr. John Owen (Edinburgh, 1850–1855), II, 34; re-issued, Grand Rapids, 1950; *Calvini opera*, XXXVIII, 75–76. This John Owen (†1867) should not be confused with the Puritan John Owen (1616–1683).

quainted with the course of the firmament see that the sky so turns. For on the north side there is a star apparent to our eye, which is as it were the axle that runs through the middle of a wheel, and the skies are seen to turn about the middle. There is another star hidden under us, which we cannot perceive, and which is called the antarctic pole. Why? Because the sky turns about that also, as though there were an axle to which the wheel was attached, as has already been said. When I speak of this motion of the heavens, I do not mean the course of the sun that we see daily, since the sun has its own special movement, whereas this is a universal motion of the whole firmament of heaven. And the said two stars are as it were fastened to those places, so that they do not move or stir. Thus you see why Job says: He stretcheth out the north . . . . The following is the reason why he says: He hangeth the earth upon nothing. On what does the earth rest? On the air. Just as we see the air above us, so is it likewise on the other side of the earth, so that the earth hangs in the middle. True, the philosophers often discuss why the earth stays this way, since it is at the very bottom of the world; and they say it is a wonder that the earth does not sink down, since nothing holds it up. Nevertheless they are able to give no other reason than is seen in the order of nature, which is so wonderful a thing.[50]

This extended discussion shows us how well Calvin absorbed the astronomy taught in his youth. Lest anyone suppose that he thought further research unnecessary, let us recall his comment on Genesis 1:16:

Astronomers . . . investigate with great labor whatever the keenness of man's intellect is able to discover. Such study is certainly not to be disapproved, nor science condemned with the insolence of some fanatics who habitually reject whatever is unknown to them. The study of astronomy not only gives pleasure but is also extremely useful. And no one can deny that it admirably reveals the wisdom of God. Therefore, clever men who expend their labor upon it are to be praised and those who have ability and leisure ought not to neglect work of that kind.[51]

Obviously it was not lack of interest in astronomy that kept Calvin from attacking Copernicus. Nor would his exegetical principles have compelled Calvin to contradict Copernicus. For, although the Genevan reformer " despised the allegorical method of interpreting Scripture, which had provided Christians with their favorite means of twisting the Bible into a religious book of their own liking," [52] he himself did not advocate or practice a narrow literalism. Thus, commenting on Ps. 136:7, Calvin asserted:

The Holy Spirit had no intention to teach astronomy; and, in proposing instruction meant to be common to the simplest and most uneducated persons, he made use by Moses and the other prophets of popular language, that none might shelter himself under the pretext of obscurity, as we will see men sometimes very readily pretend an incapacity to understand, when anything deep or recondite is submitted to their notice. Accordingly . . . the

[50] Calvin, *Sermons . . . upon the Booke of Job*, tr. Arthur Golding (London, 1574), 490 (somewhat modernized); *Calvini opera*, XXXIV, 429–430.

[51] *Calvin: Commentaries*, edd. Joseph Haroutunian and Louise Pettibone Smith (London, 1958; *Library of Christian Classics*, XXIII), 356; *Calvini opera*, XXIII, 22.

[52] Haroutunian and Smith, in *Calvin: Commentaries*, 23.

Holy Spirit would rather speak childishly than unintelligibly to the humble and unlearned.[53]

Since Calvin believed that Scripture is sometimes couched in popular language to make it intelligible to the uneducated, he need not have felt himself bound by the letter of the most explicitly anti-Copernican passage in the Bible, namely, the miracle reported in Joshua 10:12–14:

Then spake Joshua to the Lord in the day when the Lord delivered up the Amorites before the children of Israel, and he said in the sight of Israel, Sun, stand thou still upon Gibeon; and thou, Moon, in the valley of Ajalon.

And the sun stood still, and the moon stayed, until the people had avenged themselves upon their enemies. Is not this written in the book of Jasher? So the sun stood still in the midst of heaven, and hasted not to go down about a whole day.

And there was no day like that before it or after it, that the Lord hearkened unto the voice of a man.

Concerning this mightiest weapon in the Biblical armamentarium of the anti-Copernicans, Calvin might have reasoned, as he did with regard to Genesis 1:16, that the author of the Book of Joshua " described in popular style what all ordinary men without training and education perceive with their ordinary senses." [54] But Calvin did not apply this method of interpretation to Joshua's miracle. Instead, in his *Commentarius in librum Iosue*, composed during the agony of his last illness, Calvin commented:

As in kindness to the human race He divides the day from the night by the daily course of the sun, and constantly whirls the immense orb with indefatigable swiftness, so He was pleased that it should halt for a short time till the enemies of Israel were destroyed.[55]

So Calvin wrote toward the close of his life, more than two decades after the publication of Copernicus' *Revolutions*. What Calvin said about Joshua's miracle induced a nineteenth-century editor and translator of his works to remark: " One might almost suspect from this concluding sentence that Calvin was a stranger to the Copernican system." [56] This incipient suspicion will inevitably grow stronger in our minds when we recall Calvin's treatment of Ps. 19:4–6, 75:3, 93:1, 104:5, 119:90, Jeremiah 10:12, and Job 26:7. Surely this ample body of evidence authorizes us to conclude, despite Professor Shields, Canon Farrar, President White, Miss Stimson, Dean Inge, Lord Russell, Father Conway, and Dr. Will Durant, that Calvin never demolished, condemned, rejected, opposed, or stigmatized as an utter reprobate the quiet thinker who founded modern astronomy.

What, then, may we ask at the end of our inquiry, was Calvin's attitude toward Copernicus? Never having heard of him, Calvin had no attitude toward Copernicus.

---

[53] Calvin, *Commentary on . . . Psalms*, V, 184; *Calvini opera*, XXXII, 364–365.

[54] *Calvin: Commentaries*, edd. Haroutunian and Smith, 356; *Calvini opera*, XXIII, 22.

[55] Calvin, *Commentaries on the Book of Joshua*, tr. Henry Beveridge (Edinburgh, 1854), 153; re-issued, Grand Rapids, 1949; *Calvini opera*, XXV, 500.

[56] Henry Beveridge (1800–1863) in Calvin, *Commentaries on . . . Joshua*, 153.

18

The First Map to Show the Earth in Rotation

Forty years after Columbus discovered America for the Europeans, an enterprising erman Swiss publisher issued a travel book containing a map of the world. In some opies of this edition the map shows a pair of angels, one at the North Pole, the other at the outh Pole, tugging at crank handles attached to the earth's axis. By their teamwork the vo angels make our planet spin, so that this map shows the earth in rotation. It is the arliest such map.

Who made it? It bears no signature, nor is the identity of its designer certified by any utside sources, such as contemporary letters. Another intriguing question: why is it ound in some copies of this edition, but not in all the copies?

The edition was published in March, 1532, in Basel by the printer Johannes Hervagius he latinized form of Herwagen). He had received a batch of travel narratives that had een collected by Johannes Hutichius (or Huttich, to give him his German name). The job f editing the volume was entrusted to an experienced scholar, Simon Grynaeus (or runer, in German). The narratives selected as worth reprinting included such popular riters as Marco Polo, Vespucci, and Columbus. Vicente Yáñez Pinzón, who had been le captain of Columbus' Niña, reported on Brazil; Alvise de Cadomosto on West Africa; odovico di Varthema on Arabia and India. From Peter Martyr's history of the New 'orld, the volume took its title Novus Orbis. It included many other travel accounts, too umerous to be itemized here.

Since these voyages of discovery and exploration had taken Europeans far and wide to any previously unknown parts of the world, the volume was embellished by a map of the hole earth. The caption over the map reads (in Latin, like the rest of the volume) TYPUS OSMOGRAPHICUS UNIVERSALIS. This "Complete Depiction of the World" was laced right after an introductory essay entitled "Typi cosmographici et declaratio et sus" (Explanation and Use of the Depiction of the World). The title of the essay and the tle of the map both begin with the same two words (their endings differ for grammatical asons). This verbal repetition is only one of many indications that the essay and the map ere both prepared by the same man, Sebastian Münster (1489-1552), one of the greatest eographers of his time. Although his name does not appear on the map, the volume does lentify him as the author of the essay. As a professor at the University of Basel since 1529, lünster was ideally situated to help Hervagius, Huttich, and Grynaeus bring out their omposite volume entitled Novus Orbis (New World).

The New World was published in March, 1532, exactly eleven years before Nicholas opernicus founded modern astronomy by proclaiming the earth to be a planet in notion. Copernicus' earth not only rotated around its axis once a day, but also revolved round the sun once a year. In addition, Copernicus' earth was no longer located at the enter of the universe, where the earth had been stationed by his predecessors. opernicus' earth, by contrast, orbited in the third planetary slot, far removed from the niverse's center.

Did Münster agree or disagree with Copernicus? Münster's introduction to the New Vorld said: "It is the unanimous opinion of all educated people that the body of the arth ... occupies the center of the entire universe." Hence, for Münster, the earth was the central body. On Münster's view, therefore, the earth could not revolve around the center in the manner of Copernicus' earth orbiting around Copernicus' centrally located sun.

Then what about the daily rotation? For Copernicus, this was performed by the earth. Hence a point on the surface of Copernicus' earth would trace out the arctic circle as the earth rotated on its axis. On the other hand, how does the arctic circle come into existence according to Münster's introduction to the *New World*? Münster offers two alternatives:

Certain people say that the arctic circle is the circle described by the Great Bear with its forefoot. But other understand that the arctic circle is the circle described by the pole of the ecliptic during the rotation of the firs movable [or sphere of the fixed stars]. And the arctic circle is $23\frac{1}{2}°$ away from the [celestial] pole.

Whereas for Copernicus the stars were absolutely stationary, for Münster they performed the daily rotation and thereby generated the arctic circle.

On both counts, then, Münster differed with Copernicus. For Copernicus, the earth rotated and revolved. For Münster, it did neither. Then why did the map accompanying Münster's Explanation and Use of the Depiction of the World portray the earth in rotation?

We must first ask a prior question: why is the earth-rotating map missing from some copies of the *New World*? Recent discussions of this map take for granted that it formed an integral part of the *New World*. A more skeptical attitude, however, was displayed long ago by the foremost authority on early Americana. He pointed out that some copies of the *New World* lacked a map altogether, while other copies contained a modified map made for the third edition, which was published in 1555, twelve years after Copernicus' *Revolutions* in 1543. Evidently a post-Copernican map was often inserted in the still unsold stock of the pre-Copernican *New World* of 1532. Copies of that first edition of the *New World*, or of the second edition of 1537, are listed in the inventory of the books in his shop which Hervagius himself wrote in August, 1554.

The earth-rotating map did not completely conform to the ideas of Copernicus any more than it did to the thinking of Münster. True, Copernicus' earth rotated, whereas Münster's stood still. But the rotation of Copernicus' earth was not produced by any angels. On the contrary, Copernicus believed that a sphere, such as the earth, keeps on turning by itself without being affected by any outside force, whether natural or supernatural.

Whoever adorned the *New World* map with the two hard-working angels produced a fascinating compromise between traditional and forward-looking conceptions. According to an old and revered tradition, angels were responsible for the motions of the celestial bodies. But that same tradition viewed the earth neither as a celestial body nor as a moving body. On the other hand, Copernicus' earth was a celestial body that moved, but its motion was not due to any angels. Thus the *New World* map wobbles on the threshold of modern science, with one foot in the stagnant past and the other in the onrushing present.

This instability is clearly visible in the labeling of South America. This area was marked ASIA in large capital letters in the *New World*'s first edition. At that time (1532) the continent reached by the westbound voyagers was believed by some Europeans to be Asia. That name still appeared five years later in the second edition, but this time the capital letters were smaller. Then in the third and last edition of 1555 the designation "Asia" vanished altogether from the American land mass. This final version, with ASIA placed where it really belongs, is also found in some copies of the 1532 edition.

When the text of that edition was reprinted later on in the same year (on 25 October) in Paris, the Basel map was not reproduced. Instead, it was replaced by an entirely different map that had been executed by an outstanding French scientist in July, 1531. Oronce Finé's map did not in the least suggest any rotation on the part of the earth. In general, his map was obviously designed without any dependence whatever on the earth-rotating Basel map. In fact, Finé's Paris map may have been made before Münster's drawing of the Basel map was transferred to a wood block by an engraver.

This engraver, it has been suggested, was none other than the great Holbein himself. True, Hans Holbein the Younger had done a good deal of work for various Basel printers, and he had also made the illustrations for an astronomical book by Münster that was published in 1534. But by that time Holbein was no longer in Basel. He had returned there from England, and with the money he had earned by painting his magnificent portraits of English notables he had bought a house in Basel on 29 August 1528. But from that time until he went back to England in 1532, he did very little for the Basel printers. This conclusion emerged from a thorough investigation of Holbein's activities in the service of the Basel printers.

If not Holbein, who did engrave Münster's map for the *New World* of 1532? This question is answered by Münster himself. In a letter written on 29 July 1542, he says that he has been working on his German geography for nearly ten years. But he has not yet decided when to send it to press "because many of the drawings for it still have to be done, and we have lost our artist Conrad and Holbein still has not come back from England." Holbein never did come back from England, where he died in 1543. As for Conrad Schnitt, he had died in November, 1541, presumably of the plague that ravaged Basel then.

After Holbein left for England in 1532, Schnitt did most of the woodcuts needed by Basel printers until his unexpected and untimely death in 1541. (Incidentally, his surname is marvelously appropriate, since in German "Schnitt" means a cut.) An example of Schnitt's work in that period strikingly resembles what he did to Münster's *New World* map of 1532. Four years later a Basel printer named Henricpetri (Hervagius had in the meantime been obliged to leave town on account of wife trouble) published Münster's *Organum Uranicum* ("Instrument for [Observing] the Heavens"). Schnitt's illustration for *Organum*'s title page shows the celestial sphere being rotated by means of a crank. Whereas Schnitt's crank in the *New World* map of 1532 was manipulated at each end by a single handle held by only one angel, in 1536 *Organum*'s handlebars are double, requiring a total of four angels in all to keep the heavens moving.

Schnitt was no ordinary woodcutter. He was usually called, more generally, an artist. In 1530 he was elected head of his guild, and in that capacity he was a member of the Basel City Council for the next six years. In 1532, when the local university had to be reorganized on account of the Reformation, Schnitt was one of the three commissioners charged with that responsibility. For his *World Chronicle* in German, he used some sources which were then available only in the original Latin.

Hence Schnitt was undoubtedly familiar with the old debate about the function of the angels:

Do the angels move the heavenly bodies?

Has it been infallibly proved, in the opinion of some, that the angels move the heavenly bodies?

When the angels move the heavenly bodies by their own power, do they move them by the authority granted to them by God?

Can an angel move the whole mass of the earth up to the sphere of the moon?

These were among the questions discussed by a teacher and his students at the Dominican house in Venice, and sent by the general of the Dominican Order to two of his leading theologians, the archbishop of Canterbury and Thomas Aquinas, the foremost scholastic thinker. Aquinas' prompt answer to his superior, dated 2 April 1271, declared: "Not only have the philosophers proved in many ways, but the learned theologians also plainly assert" that the heavenly bodies are moved by angels.

> Proofs of this proposition abound in the books of the philosophers, who regard these proofs as rigorous. It seems to me too that it can be rigorously proved that the heavenly bodies are moved by a mind, that is, either by God directly or indirectly by means of the angels. I remember having read that not one of the saints or philosophers denied that the heavenly bodies are moved by an incorporeal creature.

"I really do not see what doubt there can be" about the proposition that the angels receive their power from God. "For I believe that nobody doubts that everything done by the angels is accomplished through the power granted to them by God."

But with regard to the proposition that "by its power an angel can move the entire mass of the earth as far as the sphere of the moon," Aquinas balked:

> It does not seem that this proposition should be asserted. . . . For not only the power of an angel but also the power of a man can make a part of the earth go up by force. But I do not believe that placing the entire element [of the earth] outside the natural order is within the power of an angel.

What is under discussion here is the possibility of displacing the earth from its (supposed) station at the center of the universe. This is not the motion accomplished by Schnitt's angels in the *New World* map of 1532. They merely make the earth rotate in place: while it spins around its own axis, it always occupies the same position in space. Could this possibility have entered the discussion of the angels and the earth in the two and a half centuries between Aquinas and Schnitt?

If the daily rotation is performed by the earth (as Copernicus insisted, and we know for a certainty), then that rotation is not performed by the heavens. They only seem to go round, while the earth seems to stand still. Actually, we go round with the earth. We do not feel that motion because it is so smooth. It was the smoothness of that motion, and their participation in it, that prevented people from understanding the true situation until Copernicus came along.

Certainly Schnitt did not understand the true situation. In the *New World* map of 1532 he had the angels turn the earth. In the *Organum* of 1536 he had the angels turn the heavens. Schnitt was neither a cosmologist nor an astronomer. He was an artist. For him, the angels were decorative features added to Münster's map and title page, respectively. His angels have no cosmic significance and no scientific meaning. They are purely aesthetic.

Those who imagine that Schnitt's *New World* angels are to be accepted as the effective agents in a rotation of the earth face a chronological difficulty: in 1532, eleven years before Copernicus' *Revolutions*, the earth was depicted in rotation. How can this pre-Copernican Copernicanism be explained? One answer is Celio Calcagnini, who wrote an essay entitled "The Heavens Stand Still, the Earth Moves, or the Perpetual Motion of the Earth."

Calcagnini's essay was printed for the first time in a partial collection of his works that was published in 1544, the year after the first edition of Copernicus' *Revolutions*. If we focus on the dates of both first printings, Calcagnini's essay is post-Copernican rather than pre-Copernican. Obviously, Schnitt's earth-rotating map of 1532 could not have been affected by the printed Calcagnini essay of 1544.

But Calcagnini's essay was published posthumously, since he died in 1541, about half a year earlier than Schnitt. Could Calcagnini's essay have circulated in handwritten form uring his lifetime? Could a manuscript copy of Calcagnini's essay have reached Schnitt in asel before 1532? When did Calcagnini write his essay?

Unfortunately he did not indicate the day on which he finished writing this essay, a ustom which he followed in some cases, but not in all. However, he dedicated this essay) Bonaventura Pistofilo, the secretary of the duke of Ferrara. Only one other essay in lat bulky volume was dedicated to Pistofilo. Fortunately, that essay is dated (3 January 525), and it was placed by the editor (the personal physician of the duke of Ferrara) nmediately after the essay on the earth's motion. That was probably written before 1525. he editor also included in the volume five letters from Calcagnini to Pistofilo. Three of lese letters bear no date. The other two are dated 8 January 1525 and 1 June 1527. The olume also includes a dialogue between Calcagnini and Pistofilo. But that too is ndated. When we try to fit these scattered bits and pieces together, they point to 1520-524 as the period during which Calcagnini wrote his essay about the motion of the earth. or from 1517 to 1520 Calcagnini was absent from Ferrara, at first on an extended iplomatic mission, and then on urgent personal business. But Calcagnini must have omposed the essay when he and Pistofilo were both in Ferrara. For in the dedication 'alcagnini acknowledges Pistofilo as the only patron of the essay from the very time when was a newborn infant and "had not yet acquired its proper birthright." Then in the essay self Calcagnini reminds Pistofilo:

I had recently told you that this vault of the heavens, which you think turns with indescribable speed, and this sun and those stars, which you believe are carried along in a single rotation, stand still and enjoy eternal quiet as they rest on their spheres.

If Calcagnini wrote his essay on the earth's motion while he was in Ferrara from 1520 to 524, did he circulate any handwritten copies? Could a copy have reached Basel in time to ffect Schnitt's *New World* map of 1532? The answer comes from Calcagnini's editor: 'alcagnini was not in the habit of circulating handwritten copies of his unpublished ritings. On the contrary, he used to toss the original drafts aside higgledy-piggledy. Had ot one of his executors patiently copied Calcagnini's writings into a volume with loving are, they would in all probability not have been preserved. Hence we may be reasonably ertain that Schnitt never saw (or even heard of) Calcagnini's essay, since it did not reach asel, where it was printed in March, 1544, until after Schnitt's death.

In Schnitt's map the earth is rotated by the angels. In Calcagnini's essay the rotation of he earth has nothing whatever to do with any angels. Instead, he propounds what may be alled a thermodynamical cause of the earth's rotation. For, according to Calcagnini,

The earth's structure has implanted in all its parts a certain extraordinary longing to enjoy the celestial light and to expose their surfaces to the rays of that eternal luminary which Almighty God placed in a lofty part of heaven so that its brilliance would illuminate everything. By its agency we differentiate daylight from darkness. By its heat the seeds of all things are germinated, nourished, and developed. It is the source of the food supplied to living things; it regulates the sequence of the seasons of the year.... For who does not see the flowers daily opening outward toward the sun, and folding their leaves together when they have been withdrawn from it for a considerable length of time, and shutting themselves up in their innermost folds as though retiring for the night?

'rom the attraction between a magnet and iron Calcagnini concludes:

Hence it is fully understood how strong the power of nature is, and how strong is that feature of it which we call "Sympathy." Accordingly, nobody now need wonder that the earth, which would surely be passive and sterile without the sun's assistance, strives so eagerly to be embraced by the sun that the earth draws from the sun the sparks of life for propagating species. Consequently, there is no reason henceforth why we should attribute any motion of the earth to an unknown deity... since nothing derived from the steady course and power of nature is rightly regarded as unnatural.

COSMOGRAPHICVS VNIVERSA

lcagnini insisted over and over again that the earth's motion was caused naturally, not
)ernaturally. This emphasis puts him poles apart from Schnitt and his angels. Even
en Calcagnini propounded a cause for the earth's rotation that had nothing to do with
: sun and thermodynamics (he shows no awareness of the discrepancy), his reasoning
nained squarely within the framework of nature:

The earth turns and twists on itself endlessly. For once it has received the impulse from nature, it can never
top. Otherwise the scheme of things would necessarily be subverted and disrupted.

For the longest time virtually everybody had thought of the earth as absolutely
.tionary. Then that entrenched dogma was challenged by two writers, one of whom was
lf a dozen years younger than the other. The older one, Copernicus, wrote his early
·rk, the *Commentariolus*, about 1510 in northeastern Europe. Some ten years later, in
rrara, Calcagnini composed his essay. Did these two men ever meet? If they did not
:et in person, did the writings of either one influence the thinking of the other?
Copernicus received the doctorate in canon law from the University of Ferrara on 31
ay 1503. Some novelists, fondly imagining themselves to be narrating the history of
.ence, have concocted a conversation between Copernicus and Calcagnini in Ferrara in
: spring of 1503. Unfortunately for this idyllic picture, Copernicus was in Ferrara only
ng enough to get his doctoral degree. He had received permission to study abroad. To
ow that he had not frittered his time away on wine, women, and song, he had to bring
·me a diploma. That cost much less in Ferrara than in the other Italian universities
1ere he had studied.
As for Calcagnini, he was not in Ferrara in 1503. On 1 February 1502 he had published
)oem (he was a poet and essayist, and not really a scientist), celebrating the marriage of
e duke of Ferrara to Lucretia Borgia. But he was then in his military phase, attached to
e quartermaster corps. He did not return to Ferrara until the autumn of 1505. By that
ne Copernicus had been back home for many months. As a matter of historical fact,
en, Copernicus and Calcagnini did not meet in Ferrara in 1503.
Nor did they ever meet anywhere else at any other time. Neither man mentions the
her in his writings. Nevertheless, it would be an astonishing coincidence if the two of
em talked about a moving earth—after so many centuries of a stationary earth—only
n years or so apart, without any mutual contact whatsoever.
Any influence of Calcagnini on Copernicus is out of the question. For Calcagnini, "the
,rth is located in the middle of the universe," a doctrine rejected by Copernicus. The
iddle of Copernicus' universe is occupied by the stationary sun, around which the earth
volves once a year. The earth's orbital motion was not so much as mentioned by
alcagnini.
Since Calcagnini did not influence Copernicus, could the effect have flowed the other
ay? Now we are on firmer ground. A copy of Copernicus' *Commentariolus* was included
y a Cracow professor in the list of his books and manuscripts which he dated 1 May 1514.
ome four years later, in April 1518, Calcagnini was in Cracow, to attend the wedding of
1 Italian lady to the Polish king (Cracow was then the capital of the kingdom of Poland).
t that time Calcagnini was made a member of the Polish nobility. If Copernicus' moving
1rth was a current topic of discussion in Cracow, and if Calcagnini heard something
bout it, our "astonishing coincidence" becomes less astonishing and less coincidental.
Copernicus' *Commentariolus* stressed the difference between appearance and reality:
1e daily rotation of the heavens is merely an appearance, due to the real rotation of the
1rth. Calcagnini's essay begins with an extended treatment of faulty perception: a
raight oar, half submerged in water, looks broken. He cites many other such stock

examples. Being essentially a literary man and certainly not an expert astronomer Calcagnini failed to absorb from the *Commentariolus* the earth's orbital revolution, it noncentrality, and other such characteristic Copernican features.

Nevertheless, the earth's daily rotation made a deep impression on Calcagnini. Whe he returned to Ferrara, he expounded this unsettling idea to his stubbornly conservativ friend Pistofilo. He wrote his essay "The Heavens Stand Still, the Earth Moves, or th Perpetual Motion of the Earth," tossed it aside, and in a perpetual motion of his own went on to other themes.

Neither Calcagnini or Copernicus had a press agent. Copernicus' *Commentariolus* di not reach Basel by 1532. Neither did Calcagnini's essay. For Münster, the earth wa stationary. The rotation of the earth was added to the *New World* map of 1532 by Schnitt That rotation was not a natural motion. It was produced by angels. The first map to sho the earth in rotation was an imaginative work of art, not a solid contribution to geograph or to science.

19

Galileo the Copernican

« The very arrangement of the bodies in the heavens, and the unified structure of the parts of the universe known to us, were in doubt until the time of Copernicus. He finally pointed out to us the true system and the true framework, in accordance with which these parts are arranged » [1].

The foregoing eulogy was uttered in Galileo's *Dialogue* of 1632 by Salviati. Of the three interlocutors in the *Dialogue*, by common consent Salviati is regarded as most often expressing the innermost convictions of the great Italian scientist, the four-hundredth anniversary of whose birth is commemorated by this volume. Despite ecclesiastical warnings and despite the consequent risks, Galileo was courageous enough to include this magnificent encomium of Copernicus in his embattled *Dialogue*.

When Salviati wants his interlocutors to look up an entry in a « Table of arcs and chords », Copernicus' book is right there at hand with the desired information [2]. By contrast, when a question arises whether some statements by an anti-Copernican author have been repeated correctly, Simplicio, the spokesman for Aristotelianism, says: « I should like to have his book, and if there were somebody who would go and get it, I would be very grateful ». Thereupon Sagredo, the wealthy Venetian in whose palace the *Dialogue* is spoken, replies: « I'll have a valet go quickly ».

---

[1] *G. G.*, VII, 480. lines 25-28. Other English renderings of this passage in Galileo's *Dialogue* are available in Thomas Salusbury's translation. as revised by G. de Santillana (G. GALILEI: *Dialogue on the great world systems*. in the Salusbury translation. revised. annotated and with an introduction by G. DE SANTILLANA, Chicago-London, 1953). p. 462, and in S. Drake's translation (G. GALILEI: *Dialogue concerning the two chief world systems — Ptolemaic and Copernican*. translated by S. DRAKE, foreword by A EINSTEIN. Berkeley – Los Angeles. 1953; revised. 1962). p. 455.

[2] *G. G.*. VII, 207, line 35; G. GALILEI ... G. DE SANTILLANA [*cit. n.* [1]]. p. 195; G. GALILEI .. S. DRAKE [*cit. n.* [1]], p. 181. The « Table » is in N. COPERNICUS: *De revolutionibus orbium coelestium*. Nuremberg. 1543, I, 12 (N. KOPERNIKUS: *Gesamtausgabe*. Munich and Berlin. 1944-1949. II. pp. 38-46).

Simplicio interjects: « He will be able to do the errand [quickly] because he will find the volume open on my desk, together with the book by that other writer who also argues against Copernicus ». Sagredo agrees: « We'll have him bring that one too, to be on the safe side .... I have sent a servant » [3]. Meanwhile the three interlocutors continue their earnest discussion, in order to avoid wasting valuable time. After a little while, Sagredo remarks: « The books have just arrived. Take them, Simplicio, and find the passage in question » [4].

We may with considerable assurance, then, assume that one of Galileo's constant companions was Copernicus' *Revolutions*. The founder of modern astronomy was admired by Galileo for his immense intellectual achievement, the first basic advance in that science in well over a thousand years. Mixed with this admiration, moreover, was a deep sympathy. This profound sense of compassion for a fellow-scientist, who too had suffered for his efforts to advance man's knowledge of the universe, found expression in what Galileo hoped would become an insertion in a revised edition of his *Dialogue*. In preparation for that eventuality (which never materialized), Galileo composed some additions. These are preserved in his own handwriting in his own copy of the original edition of 1632 (*G. G.*, VII, 10). In one of these additions, presumably composed by Galileo after his humiliating trial, conviction, and forced abjuration of Copernicanism, Sagredo exclaims, in a transparently veiled reference to Galileo himself:

« The plight of poor Copernicus seems to me deserving of pity. He can't be sure that the judgment of his teachings may not by chance fall into the hands of persons who are unable to comprehend his reasoning, which is very subtle and therefore hard to understand. These persons, being thoroughly convinced in advance by ... empty appearances that Copernicus' teachings are wrong, go about pronouncing them to be false and erroneous » (*G. G.*, VII, 357, lines 12-18).

The addition containing the foregoing passage was not translated by de Santillana. Instead he supplied a short summary, in which he said that « Sagredo ... is fearful of the fate of Copernicus » [5]. However, the addition shows Sagredo, not « fearful of the fate of Copernicus », but full of pity for his plight (mi par degna di commiserazione la condizione del povero Copernico).

---

[3] *G. G.*, VII, 247, line 34-248, line 6; G. GALILEI ... G. DE SANTILLANA [*cit. n.* [1]], pp. 236-237; G. GALILEI ... S. DRAKE [*cit. n.* [1]], p. 221.

[4] *G. G.*, VII, 257, lines 29-30; G. GALILEI ... G. DE SANTILLANA [*cit. n.* [1]], p. 247; G. GALILEI ... S. DRAKE [*cit. n.* [1]], p. 231.

[5] G. GALILEI ... G. DE SANTILLANA [*cit. n.* [1]], p. 342.

Whereas de Santillana merely mistook Sagredo's (i. e., Galileo's) attitude regarding Copernicus as fear rather than pity, Drake's version makes Galileo say that Copernicus « could expect only censure for his views » [6]. If so, why would Copernicus have finally permitted his *Revolutions* to be published, after prudently withholding that revolutionary book from the printer for so many decades? Is it true that Copernicus « could expect only censure for his views », after his trial balloon (Rheticus' *First report* [7]) was sent aloft twice without drawing down any lightning on the heads of the Copernicans? Would Galileo, an adroit and experienced skirmisher, have expressed unqualified admiration for Copernicus, if the latter had published views for which « he could expect only censure »? Is this what Galileo really said about Copernicus in the addition? A glance at the Italian text reveals that Galileo said no such thing. What the addition says about Copernicus is that « non si può tener sicuro che la censura delle sue dottrine non possa per avventura cadere in mano di persone » etc. These words do not mean that Copernicus « could expect only censure for his views ». These words do mean that Copernicus « can't be sure that the judgment of his teachings may not by chance fall into the hands of persons » unqualified to act as impartial judges.

Drake's version of the addition, as it proceeds, makes Galileo say that Copernicus « could not let them [his views] fall into the hands of anyone who ... would be convinced of their falsity ». But if an author consents to the publication of his views, how can he prevent them from falling into the hands of anyone who would be convinced of their falsity? Neither Copernicus nor Galileo believed that he could prevent such a situation from occurring. In fact, both of them were acutely aware that such a situation was highly likely. In sober truth, Galileo's addition made no such statement as is assigned to it by Drake's version. The statement made by Galileo's addition is that Copernicus « can't be sure that the judgment of his teachings may not by chance fall into the hands of persons who » are « thoroughly convinced in advance ... of their falsity ».

Such a person, in Drake's version, « would be convinced of their falsity on account of some superficial appearances ». But Galileo does not say « would be convinced ». On the contrary, he says « thoroughly convinced in advance » (« ben di già persuasi »).

Drake's omission of « di già », together with his misconstruction of the sentence, completely alters its meaning, and thereby utterly distorts Galileo's attitude toward Copernicus. Is this distortion an unintended

---

[6] G. Galilei ... S. Drake [*cit. n.* [1]], p. 329.

[7] G. J. Rheticus: *Narratio prima*, Danzig, 1540; translated into English by E. Rosen: *Three Copernican treatises*. New York - London. 1959 (2nd ed.). pp. 107-196.

by-product of Drake's « taking certain liberties with the text » [8]? Should « Galileo's known abhorrence of pedantry » be invoked to condone such gross errors?

It will be recalled that a servant brought the interlocutors two anti-Copernican books. One of these, says Simplicio, ridicules the annual motion ascribed to the earth by Copernicus; for if the earth revolves in an orbit, « the sun, Venus, and Mercury are lower than the earth; ... Christ, our Lord and Redeemer, went up to hell, and went down to heaven when he approached the sun; and when Joshua commanded the sun to stand still, the earth stopped ». Salviati (i.e., Galileo) comments:

« I am satisfied with everything, except that he mixed up passages from the ever venerable and awe-inspiring Sacred Scriptures with these regrettably insolent puerilities, and that he sought to assail with holy things a man who, philosophizing for the fun of it and in sport, neither affirms nor denies, but reasons informally after making some assumptions or hypotheses » [9].

Such misuse of the *Bible* had been prophetically foreseen by Copernicus, whose long delay in publishing his masterpiece was partly caused by his « fear that the theologians would contradict » him [10]. And in dedicating his *Revolutions* to Pope Paul III, Copernicus had said:

« Perhaps there will be prattlers (mataiologoi) who, although completely ignorant of mathematics, nevertheless take it upon themselves to pass judgment on mathematical questions, and on account of some passage in Scripture, badly distorted to their purpose, will dare to censure and assail what I have done here » [11].

This portion of Copernicus' dedication was left untranslated in the original Latin when it was quoted by Galileo in his Italian *Letter to the Grand Duchess* (*G. G.*, V, 314, lines 3-6). When this eloquent plea for the liberation of science from theological shackles was recently translated into English, « mataiologoi », the Greek word used by Copernicus for « prattlers », was misequated with « exegetes » [12]. This version imputes to Copernicus the belief that exegetes — specialists in interpreting the *Bible* — might badly distort some passage in Scripture. Did any such thought ever enter the mind of Copernicus, a canon of the Roman Catholic

---

[8] G. GALILEI ... S. DRAKE [cit. n. 1], p. xxv.

[9] *G. G.*, VII, 384, lines 21-25, 32-36; G. GALILEI ... G. DE SANTILLANA [cit. n. 1], p. 367; G. GALILEI ... S. DRAKE [cit. n. 1], p. 357.

[10] J. KEPLER: *Opera omnia*, ed. CH. FRISCH, Frankfurt a. M. - Erlangen. 1858-1871. I. p. 246, line 17.

[11] N. KOPERNIKUS: [cit. n. 2]. II. p. 6. lines 27-30; *Paul's Epistle to Titus*. 1:10.

[12] S. DRAKE: *Discoveries and opinions of Galileo*. Garden City. 1957. p. 180.

church? Even if it did, was he not much too discreet to suggest any such possibility in a dedication addressed to the reigning pope?

Copernicus believed that there might be prattlers (not exegetes) who would « dare to censure and assail what I have done here » (meum hoc institutum). These last three words were englished as « this hypothesis of mine » [13]. But whatever Copernicus may have meant by « institutum », he certainly did not mean « hypothesis ». To depict Copernicus as labeling his life's work a « hypothesis », and as seeking a quarrel with exegetes, is to misunderstand both his scientific outlook and his social behavior.

The first of these two misunderstandings is nothing new. Thus Salviati's comment, quoted above, refers to « a man who, philosophizing for the fun of it and in sport, neither affirms nor denies, but reasons informally after making some assumptions or hypotheses ("ipotesi") ». To be sure, there is a vast difference between an outsider stating that a man makes some hypotheses along the way, and the man himself calling his own finished product « this hypothesis of mine ». Moreover, Salviati (i.e., Galileo) here carefully refrains from mentioning Copernicus by name, although within its own context the reference to « a man who, philosophizing for the fun of it ... reasons informally after making some ... hypotheses » is unmistakably intended as a characterization of Copernicus [14].

This was how Copernicus was characterized by Galileo after he lost the battle of 1616, when the Sacred Congregation of the Index suspended Copernicus' *Revolutions* « until corrected » [15]. But on the eve of that battle, in a private letter to a loyal friend in Rome, he zealously opposed any official pronouncement which would « declare that Copernicus did not hold the earth's motion to be really a physical fact; that on the contrary, as an astronomer, he adopted it merely as a hypothesis which, although false in itself, was convenient in accounting for the observed phenomena; and that its use as such was therefore to be permitted, and belief in its truth prohibited ». « This », Galileo correctly concluded, « would be tantamount to declaring that that book had not been read ». With regard to his own attitude, Galileo informed his correspondent, he did not want authorities in Rome to believe that he « approved Coper-

---

[13] S. DRAKE: [cit. n. 12], p. 180.

[14] Emil Strauss, who translated the *Dialogue* into German (G. GALILEI: *Dialog über die beiden hauptsächlichsten Weltsysteme, das ptolemäische und das kopernikanische*, übersetzt und erläutert von E. STRAUSS, Leipzig, 1891, p. 558) mistakenly transferred this characterization of Copernicus to the anti-Copernican author.

[15] Latin text of the Index decree of March 5, 1616 in VON GEBLER: *Galileo Galilei and the Roman Curia*, English translation by G. STURGE, London, 1879, pp. 345-346; at p. 84, the decree in English translation (reproduced by G. DE SANTILLANA, *The crime of Galileo*, Chicago, 1955, p. 123).

nicus' position only as an astronomical hypothesis which in fact, however, is not true » (*G. G.*, XII, 183-185, lines 36-41, 44-46).

According to Galileo in 1615, only those who had not read Copernicus could view his distinctive doctrine as a convenient hypothesis, false in fact. Yet Galileo's *Dialogue* of 1632 depicted Copernicus, not as seriously seeking the whole truth and finding a sizable chunk of it, but as « philoso-phizing for the fun of it », as being a man who « neither affirms nor denies, but reasons informally ». This contrast between Galileo's two portraits of Copernicus (the earnest thinker of 1615, the affable entertainer of 1632) is indeed striking. Yet Galileo's motive in so completely altering his image of Copernicus will be entirely clear to those who are familiar with what the authorities in Rome did to him in 1616. He did not belong to the exalted group who discussed Copernicus without having read the *Revolu-tions*. On the contrary, as we have already seen good reason to believe, that book was one of his constant companions. From it he obtained the genuine pigments with which he painted his authentic portrait of Coper-nicus in 1615. The tarnished caricature of 1632, on the other hand, was smeared on rotting paper with the spurious materials supplied by the purveyors of coerced opinion.

What we are confronted with here, then, is altogether different from Galileo's five errors concerning Copernicus in his *Letter to the Grand Duchess* [16]. There he made a few honest mistakes, whereas here he con-sciously falsifies, for prudential reasons of self-protection.

It may perhaps be objected that « conscious falsification » is an un-deserved epithet for a description of Copernicus as a man who enunciated hypothetical views. For does not Copernicus himself in the *Revolutions* often call the earth's motion a « hypothesis »? To be sure, but there the term « hypothesis » does not mean a tentative conjecture or uncertain supposition, and certainly not a known falsehood. On the contrary, Coper-nicus used « hypothesis » in its strictly etymological sense to denote an underlying proposition, basic to a theoretical system [17].

To declare that Copernicus used the term « hypothesis » in its non-etymological sense would be tantamount, we heard Galileo exclaim in 1615, to declaring that the *Revolutions* had not been read. Those non-readers recently acquired a defender, who tried to turn the point of Galileo's sword against the author of the *Dialogue* himself. « Not even Galileo seems to have read » Copernicus' *Revolutions*, we are told; Galileo « had never taken much interest in the tiresome details of planetary theory, and there was no real reason for him to plod through the technical chapters in the *Revolutions* from cover to cover. If he had done so, he could not

---

[16] E. ROSEN: *Galileo's misstatements about Copernicus*, below, 193–204.

[17] E. ROSEN: [*cit. n.* 7], pp. 28-33.

have ... attributed the idea to Copernicus that the moon either shines in her own light or is transparent to the light of the sun»[18]. These alternatives were mentioned by Galileo in connection with Venus, not the moon (*G. G.*, VII, 362, lines 8-10). Moreover, they were mentioned by Copernicus [19], not in his «technical chapters», devoted to «the tiresome details of planetary theory», but in his «broad outline», where «the basic principles and the program of the work are all set out in the first eleven chapters of the first book»[20]. Evidently it takes a non-reader to defend non-readers.

Our modern non-reader calls Copernicus' *Revolutions* «The Book That Nobody Read»[21]. If nobody read the first edition (Nuremberg, 1543), why would an astute publisher have invested good money in issuing a second edition in 1566? This venture was undertaken by a Basel businessman, whose publishing career extended over fifty years, and whose attendance at 108 Frankfurt book fairs [22] would indicate that he could gauge sixteenth-century readers' wishes more accurately than our modern non-reader can now. The folio format of the first two editions was somewhat reduced by the publisher of the third edition (Amsterdam, 1617), so that it could be bound with a set of tables previously issued by the editor, a Dutch professor of mathematics, «in order that in this way students of astronomy may have a work absolutely complete, in practice as well as in theory»[23]. Three-quarters of a century, then, after the first edition of Copernicus' *Revolutions*, «The Book That Nobody Read» was still being utilized professionally.

Our non-reader lumps the first three editions of the *Revolutions* with the fourth (Warsaw, 1854) and fifth (Thorn, 1873), as though these five editions all belonged in the same pigeonhole. He gives no indication that he is aware of the fundamental difference between these two groups of editions. Whereas the first three editions had been workaday manuals for practical astronomers in all countries, the Warsaw and Thorn editions were designed as rival monuments to Polish and German national pride. Whereas the first three editions were bunched together in three-quarters of a century (1543-1617), it was more than two centuries later before the national monuments began to appear (1617-1854).

---

[18] A. KOESTLER: *The sleepwalkers*. London - New York. 1959, pp. 192, 479.

[19] N. COPERNICUS: [*cit. n.* [2]]. I. 10; N. KOPERNIKUS: [*cit. n.* [2]]. II. pp. 22. line 35-23. line 1.

[20] A. KOESTLER: [*cit. n.* [18]]. p. 193.

[21] A. KOESTLER: [*cit. n.* [18]], p. 191.

[22] I. STOCKMEYER and B. REBER: *Beiträge zur Basler Buchdruckergeschichte*. Basel. 1840. pp. 147-150.

[23] N. COPERNICUS: *Astronomia instaurata*. Amsterdam. 1617. *verso* of the title-page.

In those two centuries Copernicus' *Revolutions* became an honored classic of science because it had ceased to be a practical manual. It ceased to be a practical manual, because it was superseded by subsequent masterpieces, such as Kepler's *Epitome of Copernican astronomy*, Galileo's *Dialogue*, and Newton's *Mathematical principles of natural philosophy*. It was the great Copernicans, including Galileo, who preserved the solid underpinning of the *Revolutions*, discarded its extraneous trimmings, and added the new wings which completed the splendid structure of the Copernican cosmology. The long, slow process of construction was started by Copernicus' *Revolutions*, which was consulted by the later builders until it became obsolete. But for a considerable time it was indispensable to those who, like Galileo, endeavored to improve man's knowledge of the physical universe.

Our non-reader tells us that Galileo's *Sunspots* contained « his first printed statement in favor of the Copernican system. Up to this date — we are now in 1613, and he is nearly fifty — he had defended Copernicus in conversations at dinner tables, but never in print .... Here it was at last, the first public commitment, though somewhat vague in form, a full quarter-century after Kepler had first sounded the Copernican trumpet in the *Mysterium* » [24].

Kepler's *Mysterium* was published in 1596, eight full years less than « a full quarter-century » before Galileo's *Sunspots*. Five years before « Kepler had first sounded the Copernican trumpet in the *Mysterium* », that ill-fated genius Giordano Bruno had published his stirring paean in praise of Copernicus in a work which Kepler knew well [25]. There is nothing vague about the form of the commitment to Copernicanism in Galileo's *Sunspots*:

« And perhaps this planet also [Saturn], no less than horned Venus, harmonizes admirably with the great Copernican system, to the universal revelation of which doctrine propitious breezes are now seen to be directed toward us, leaving little fear of clouds or crosswinds » [26].

This commitment is « somewhat vague in form », we are told here, and later our non-reader remarks:

« The *Letters on sunspots* were the only printed work by Galileo which contained a favorable reference to the Copernican system; but ... that reference treated it merely as a hypothesis » [27].

[24] A. KOESTLER: [*cit. n.* [18]], pp. 430-431.

[25] G. BRUNO: *De immenso*, Frankfurt, 1591, pp. 327-336; G. BRUNO: *Opera latine conscripta*, Naples and Florence, 1879-1891 (facsimile reprint, Stuttgart-Bad Cannstatt, 1962), I, pp. 380-389.

[26] *G. G.*, V, 238, lines 26-31; A. KOESTLER: [*cit. n.* [18]], pp. 430-431.

[27] A. KOESTLER: [*cit. n.* [18]], p. 458.

Does « the great Copernican system » (« gran sistema Copernicano ») sound like a mere hypothesis? What is « vague in form » about « universal revelation ... leaving little fear of clouds or crosswinds » (« palesamento universale ... poco ci resta da temere tenebre o traversie »)?

Our non-reader assures us that the « *Letters on sunspots* were the only printed work by Galileo which contained a favorable reference to the Copernican system ». Our non-reader evidently failed to read his own re-publication of the Sentence terminating the Inquisition's trial of Galileo in 1633:

« Through the publication of the said book [Galileo's *Dialogue* of 1632] the false opinion of the motion of the earth and the stability of the sun was daily gaining ground .... In this book you have defended the said opinion » [28].

Our non-reader evidently failed to read his own re-publication of the Abjuration squeezed out of Galileo by the Inquisition:

« I wrote and printed a book [the *Dialogue*] in which I discuss this new doctrine ... and adduce arguments of great cogency in its favor » [29].

Our non-reader tells us that Galileo's « first explicit public pronouncement in favor of the Copernican system was only made in 1613 » [30]. Yet our non-reader himself quoted from Galileo's *Sidereal message* of 1610 the following explicit public pronouncement in favor of the Copernican system:

« We have an excellent and exceedingly clear argument to put at rest the scruples of those who can tolerate the revolution of the planets about the sun in the Copernican system, but are so disturbed by the revolution of the single moon around the earth while both of them describe an annual orbit round the sun, that they consider this theory of the universe to be impossible » [31].

The above statement in Galileo's *Message* in favor of the Copernican system elicits the following comment from our non-reader:

« In other words, Galileo thought the main argument of the anti-Copernicans to be the impossibility of the moon's composite motion around the earth, and with the earth around the sun » [32].

These are not only « other words », but also other ideas. For, we may ask, where does Galileo say that he thought the moon's composite motion

[28] A. KOESTLER: [*cit. n.* 18], p. 604; *G. G.*, XIX, 404, lines 60-64.

[29] A. KOESTLER: [*cit. n.* 18], p. 607; *G. G.*, XIX, 406, lines 161-163.

[30] A. KOESTLER: [*cit. n.* 18], p. 357; cfr. *note* 24, above.

[31] A. KOESTLER: [*cit. n.* 18], p. 366.

[32] A. KOESTLER: [*cit. n.* 18], p. 366.

to be anybody's « main argument »? Furthermore, are « the anti-Copernicans » as a whole to be identified with « those who can tolerate the revolution of the planets about the sun in the Copernican system »? Were there not many more anti-Copernicans who could not tolerate the revolution of the planets about the sun?

According to our non-reader, the remark in Galileo's *Message* about the moon's composite motion « was the only reference to Copernicus in the whole booklet ». Yet in the dedication of the *Message* Galileo referred to « the center of the universe, namely, the Sun » (*G. G.*, III, 56, lines 14-15). He thereby publicly proclaimed his adherence to the Copernican system which, of the then competing cosmologies, was the only one to put the sun in the center of the universe. Another tenet then upheld exclusively by the Copernican system was affirmed by the following passage in Galileo's *Message*:

« To those who exclaim that the Earth must be excluded from the ballet of the heavenly bodies, mainly for the reason that it has no motion and light, I shall prove ... that it moves and surpasses the Moon in brilliance » (*G. G.* III, 75, lines 10-13).

So convincing was the pro-Copernican argumentation in Galileo's *Message* that the publisher of a well-known *Defence of Galileo* attributed to Kepler the statement that « immediately after Galileo's *Sidereal message* most philosophers wanted to be Copernicans » [33]. As a fellow-Copernican, Kepler may have been inclined to overestimate the persuasive effect of Galileo's *Message* on philosophers and scientists in 1610 and thereafter. But even if this be an overestimate, it is at least the considered judgment of a well-informed contemporary, who did not share our modern non-reader's notion that Galileo's *Message* contained only one reference to Copernicus, and « no explicit commitment » to Copernicanism.

Another notion befuddling our non-reader is that Copernicus' *Revolutions* « remained on the Index for exactly four years » [34]. If the *Revolutions* remained on the Index only from 1616 to 1620, why did Leibniz in 1694 say that a little paper which he had sent to a leading Italian mathematician « seemed to me the right thing to persuade the gentlemen in Rome to permit Copernicus' opinion » [35]? Why do we find Copernicus' *Revolutions* still listed in the *Index of prohibited books*, published at Rome by the press of the Apostolic Camera, by order of pope Pius VII, in 1819 [36]?

---

[33] T. Campanella: *Apologia pro Galileo*, Frankfurt, 1622, p. 4.

[34] A. Koestler; [*cit. n.* [18]], p. 458.

[35] G. W. Leibniz: *Mathematische Schriften*, ed. C. J. Gerhardt, Berlin-Halle, 1849-1863 (re-issued, Hildesheim, 1962), II, 199; cfr. VI, pp. 144-147.

[36] *Index librorum prohibitorum*, Rome, 1819, p. 77.

Why not frankly admit that Copernicus' *Revolutions* remained on the *Index* for over two centuries, a somewhat longer period than « exactly four years »?

We do not know exactly when Galileo became a Copernican. But in a private letter thanking Kepler for sending him a copy of the *Cosmographic mystery*, Galileo wrote in 1597:

« Many years ago I accepted Copernicus' theory, and from that point of view I discovered the reasons for numerous natural phenomena, which unquestionably cannot be explained by the conventional cosmology. I have written down many [supporting] arguments as well as refutations of objections. These, however, I have not dared to publish up to now. For I am thoroughly frightened by what happened to our master, Copernicus. Although he won immortal fame among some persons, nevertheless among countless [others] — for so large is the number of fools — he became a target for ridicule and derision. I would of course have the courage to make my thoughts public, if there were more people like you. But since there aren't, I shall avoid this kind of activity » (*G. G.*, X, 68, lines 17-27).

If we wish to understand why Galileo avoided this kind of activity — publishing his pro-Copernican manuscript — until he was well past the midpoint of his career, we should look at the reactions of some of those who read « The Book That Nobody Read ».

Kepler tells us: « When I entered upon the study of philosophy at the age of eighteen in 1589, the young fellows were busy with Julius Caesar Scaliger's *Exoteric exercises* » [37]. Another young fellow who busied himself with this immensely influential book was Galileo, who cited one of its arguments in a memorandum connected with a work which he wrote in 1589 and shortly thereafter (*G. G.*, I, 412, lines 1-2). Presumably he was familiar, therefore, with J. C. Scaliger's recommendation that certain « writings should be expunged or their authors whipped » [38]. This amiable recommendation by one of the intellectual giants of the sixteenth century was aimed directly at Copernicus by a marginal note. Scaliger's recommendation was echoed by another prominent opinion-maker, Francesco Maurolico, Sicily's second Archimedes. This illustrious mathematician declared that « Nicholas Copernicus ... deserves a whip or a scourge rather than a refutation » [39]. With specific reference to Copernicus, Philip Melanchthon, the preceptor of Germany, urged that « wise governments ought

---

[37] J. KEPLER: [*cit. n.* [10]]. I, p. 2. lines 19-21; VIII. p. 673. lines 24-25.

[38] J. C. SCALIGER: *Exotericarum exercitationum liber*, Paris. 1557, c. 142ʳ.

[39] F. MAUROLICO: *Opuscula mathematica*. Venice. 1575. p. 26: see E. ROSEN: *Maurolico's attitude toward Copernicus*. in *Proceedings of the American Philosophical Society*. 101 (1957). pp. 177-194.

to repress impudence of mind » [40]. Many of Copernicus' opponents called his theory absurd, and the astronomer himself insane, in a period not especially renowned for its humane treatment of the demented.

These denunciations did no physical harm to Copernicus himself. So persistently had he resisted the efforts of his well-intentioned but short-sighted supporters to persuade him to publish the *Revolutions* that when it finally was printed, his impudent mind could no longer be repressed by wise governments, nor his body whipped by lashes raised at the behest of the foremost molders of opinion.

In this harsh climate of opinion Galileo chose not to publish his pro-Copernican manuscript. He did not relish the prospect of having his impudent mind repressed by wise governments, nor his body whipped or committed to an insane asylum. Our non-reader « still wonders at the motives of his secrecy » [41]. Wisdom is said to begin with wonder.

Galileo's determination to keep his Copernicanism hidden from public view was undermined when unexpected ocular evidence spectacularly strengthened Copernicus' mathematical and astronomical reasoning. Galileo's electrifying discoveries with the telescope, by which he inaugurated a new period in the history of observational astronomy, were announced in his *Message* of 1610. In that little bombshell he first disclosed his acceptance of Copernicanism. Mounting clerical opposition induced him, as a faithful Roman Catholic, to try to dissuade his church from confining itself too narrowly within an obsolescent cosmology. This antiquated edifice was beginning to disintegrate under the battering it received from every fresh advance in science. The cracks in the crumbling building could not be filled by a condemnation of Copernicanism. But the forces arrayed against Galileo's loyal and valiant effort overpowered him. In the end he too was condemned. His condemnation « represents », in the words of the greatest Galileo scholar of all times, « if not the greatest error, one of the biggest errors of the Roman Curia » [42].

---

[40] *Alter libellus epistolarum Philippi Melanthonis*. Wittenberg. 1570 (reprinted. 1574). pp. 334-335; *Corpus reformatorum*. Halle. 1837. IV. p. 679.

[41] A. KOESTLER: [*cit. n.* [18]]. p. 357.

[42] A. FAVARO; *Galileo e l'inquisizione*. Florence. 1907. p. 9.

20

Galileo's Misstatements about Copernicus

A RECENT English translation [1] of selections from the writings of Galileo (1564–1642) will doubtless bring to the attention of many readers the statements about Copernicus (1473–1543) in the great Italian scientist's *Letter to the Grand Duchess Christina*. These statements by Galileo contain five serious historical errors. To impede their further spread is the aim of the present article.

The first of the five errors occurs in Galileo's remark that "Nicholas Copernicus . . . was not only a Catholic, but a priest and a canon." [2] In a preliminary formulation [3] he had said: "Nicholas Copernicus was not only a Catholic, but a member of the regular clergy and a canon." [4] In both these versions,

[1] *Discoveries and opinions of Galileo*, translated by Stillman Drake (New York: Doubleday, 1957); reviewed in the *Journal of the History of Ideas*, 1957, *18*: 439–448, by Edward Rosen, and in *Isis*, 1957, *48*: 378–379, by Giorgio de Santillana.

[2] Drake, p. 178; *Le opere di Galileo Galilei*, national edition (Florence, 1890–1909; reprinted 1929–1939; cited hereafter as "NE"), V, 312.4–6: "Niccolò Copernico fu . . . uomo non solamente cattolico, ma sacerdote e canonico."

[3] Galileo's letter of 16 February 1615 to his good friend Piero Dini, who was then an official at the papal court, and a few years later became an archbishop. Demetrio Marzi (1862–1920), La questione della riforma del calendario nel quinto concilio lateranense (1512–1517), *Pubblicazioni del r. istituto di studi superiori in Firenze*, sez. di filosofia, 1896, *27*: 218, said that in Galileo's *Letter to the Grand Duchess* in his letter to Dini there were "some minor errors" (qualche piccola inesattezza), without specifying what these minor errors were. Marzi himself (p. 217) committed the minor error of misdating Galileo's letter to Dini "6 February 1614," even though he cited NE, which gives the date of the letter correctly as 16 February 1615. The minor error of the date led Marzi into a major error concerning the chronological relationship between the letter to Dini and the *Letter to the Grand Duchess*; according to Marzi (pp. 217–218), Galileo wrote the *Letter to the Grand Duchess* the year following ("l'anno seguente") his letter to Dini. Yet in the letter to Dini Galileo explained "what a pernicious thing it would be to proclaim as doctrine settled by Holy Scripture any propositions whose contrary may some day be demonstrated; with regard to these matters I have written a very extensive discussion, which is not yet in good enough condition for me to send you a copy, although I shall do so as soon as possible" (NE, V, 292.20–24). Hence, despite Marzi, Galileo did not write the *Letter to the Grand Duchess*

in the year following his letter to Dini. In that letter Galileo described the *Letter to the Grand Duchess* as already written, lacking only the final touches ("l'ultima mano"; NE, XII, 181.8). Evidently Marzi forgot that "Di Firenze, li 16 Febbraio 1614" (NE, V, 294.18), the close of Galileo's letter to Dini, followed "the Florentine style which, as is known, from January to 24 March was a year behind the present modern style" (Marzi, p. 30, n. 2). Marzi himself (p. 124, n. 4) pointed out that a book dated 10 January 1514 by its Florentine publisher was actually issued, according to the modern style, in 1515 (cf. Marzi, p. 142, n. 1).

[4] NE, V, 293.9–10: "Niccolò Copernico fu uomo non pur cattolico, ma religioso e canonico." If the word "e" is omitted from this sentence, "religioso" is transformed from a substantive into an adjective. As a substantive, "religioso" refers to a member of a monastic order, but as an adjective it merely means "pious." Hence the omission of "e" would cancel Galileo's description of Copernicus as a member of the regular clergy. This description is indeed missing in Emil Wohlwill (1835–1912), *Galilei und sein Kampf für die Copernicanische Lehre* (Hamburg and Leipzig, 1909–1926), I, 522, where Wohlwill's paraphrase of Galileo's letter to Dini has Galileo say: "Copernicus was not only a Catholic, but also a pious canon" (ein frommer Kanonikus), without any mention of his belonging to a religious order. Although Wohlwill always cited NE in the published version of his book, he may actually have read Galileo's letter to Dini in an earlier edition which omitted the "e" (*Le opere di Galileo Galilei*, Florence, 1842–1856, ed. Eugenio Albèri, II, 15). Albèri took the text of the letter (with "e" omitted) from Giambatista Venturi, *Memorie e lettere inedite finora o disperse di Galileo Galilei* (Modena, 1818–1821, I, 209). Venturi in turn had obtained the text from Jacopo Morelli, who printed the letter for the first time (*I codici manoscritti volgari della libreria Naniana*, Venice, 1776, p. 193). Morelli had found a copy of the letter (the original in Galileo's own handwriting has not survived) in the collection of manuscripts he was describing for publication; twenty years later the

as the attentive reader will have noticed, Galileo characterized Copernicus as a Catholic and a canon. His preliminary description of Copernicus as a member of the regular clergy, however, was not repeated by Galileo in his *Letter to the Grand Duchess*. Does not his failure to reiterate the claim that Copernicus belonged to a religious order signify a realization on Galileo's part that he could not substantiate this claim? Nor is the situation any better with regard to Galileo's assertion that Copernicus was a priest. No evidence that Copernicus entered the priesthood was known to Galileo. In fact, it was more than three centuries after he composed his *Letter to the Grand Duchess* before any document allegedly designating Copernicus as a priest was published.[5] Although this alleged designation has been accepted by scholars too numerous to be listed here, it is nevertheless historically worthless, as I shall undertake to demonstrate on another occasion.[6] The simple truth of the matter is that Copernicus was neither a monk nor a friar nor a priest.

In order to perceive Galileo's second error, let us resume reading his *Letter to the Grand Duchess* at the point where our quotation from it stopped. Galileo continues: Copernicus was "so esteemed by the church that when the Lateran Council under Leo X took up the correction of the church calendar, Copernicus was called to Rome from the most remote parts of Germany to undertake its reform."[7] But Copernicus does not say that he was called to Rome. He does say that

> . . . not so long ago under Leo X the Lateran Council considered the question of reforming the ecclesiastical calendar. The problem remained unresolved then only because it was felt that the lengths of the year and month and the motions of the sun and moon had not yet been adequately measured. From that time on I have directed my attention to a closer study of these topics, at the instigation of that most distinguished man, Paul, bishop of Fossombrone, who was then in charge of this matter.[8]

collection was willed by its owner, Jacopo Nani, to the Biblioteca Marciana in Venice, where the letter is now catalogued as no. 5547 (formerly It. IV, 59–60; see Carlo Frati and Arnaldo Segarizzi, *Catalogo dei codici marciani italiani*, Modena, 1909–1911, II, 45). In Nani's MS (and Morelli's edition) Galileo's letter to Dini contained the "e." Venturi dropped the "e," not by the exercise of superior editorial judgment, but by sheer inadvertence. Venturi's careless omission of the "e" was uncritically followed by Albèri, who was followed equally uncritically by Wohlwill. Although the latter cited NE, in this instance he did not consult it. For after comparing five MSS (NE, V, 270–271), NE restored the correct reading of the first editor, Morelli, and of the second editor, who utilized a Florentine MS and likewise printed the "e" (Giovanni Targioni Tozzetti, *Notizie degli aggrandimenti delle scienze fisiche accaduti in Toscana nel corso di anni LX del secolo XVII = Atti e memorie inedite dell' Accademia del Cimento,* Florence, 1780, II, 28).

[5] Lino Sighinolfi, *Domenico Maria Novara e Nicolò Copernico, Studi e memorie per la storia dell' Università di Bologna,* 1920, 5: 216, 232.

[6] Edward Rosen, *Copernicus was not a priest* (forthcoming). This article will document the remark made by the present writer in an address delivered at the Copernicus Quadricentennial Celebration in Carnegie Hall on 24 May 1943 and published in *Nicholas Copernicus, a tribute of nations,* ed. Stephen P. Mizwa (New York: Kosciuszko Foundation, 1945), p. 30: "It is sometimes erroneously stated that Copernicus

became a priest or a monk; but as a matter of fact he never took holy orders and he never joined any of the regular monastic brotherhoods."

[7] Drake, p. 178; NE, V, 312.6–9: "tanto stimato, che, trattandosi nel concilio Lateranense, sotto Leon X, della emendazion del calendario ecclesiastico, egli fu chiamato a Roma sin dall' ultime parti di Germania per questa riforma." If we compare this Italian text with Drake's translation, we see that the words "by the church," which we have no counterpart in Galileo, were inserted by Drake, who himself labeled (p. vii) his own translations free, rather than precise. Does not his interpolation of the three words "by the church" significantly alter Galileo's meaning? By calling Copernicus "tanto stimato," surely Galileo meant that Copernicus was held in high esteem generally, and not merely by the church.

[8] *De revolutionibus orbium coelestium* (Nuremberg, 1543, fol. 4v), Dedication, near the end: "non ita multo ante sub Leone X. cum in Concilio Lateranensi vertabatur quaestio de emendando Calendario Ecclesiastico, quae tum indecisa hanc solummodo ob causam mansit, quod annorum et mensium magnitudines, atque Solis et Lunae motus nondum satis dimensi haberentur. Ex quo equidem tempore, his accuratius observandis, animum intendi, admonitus a praeclarissimo viro D. Paulo episcopo Semproniensi, qui tum isti negotio praeerat." The last five words do not mean "who had been present at those deliberations," despite Charles Glenn Wallis, in *Great books of the western*

While the Fifth Lateran Council (1512–1517) was in session, Pope Leo X announced that he had "consulted the greatest experts in theology and astronomy," [9] whom he had "advised and encouraged to think about remedying and suitably correcting" [10] the calendar. He added that "they have conscientiously heeded me and my instructions, some of them in writing, others orally." [11] But when these written and oral discussions produced no suitable correction, Leo X issued a general appeal. To the Holy Roman Emperor, for example, he dispatched a message urging that "of all the theologians and astronomers whom you have in your empire and domains, you should order . . . every single one of high renown . . . to come to this sacred Lateran Council. . . . But if there be any who for a legitimate reason cannot come to the Council, Your Majesty will please instruct them . . . to send me their opinions carefully written." [12] A similar notice was distributed in printed form to the heads of other governments and of all universities. [13] Apart from this general invitation, which was twice repeated, [14] Copernicus received no special call to Rome, despite Galileo's misstatement to that effect.

The experts originally consulted by Leo X replied, it will be remembered, "some of them in writing, others orally." In like manner, those for whom the later appeal was intended were ordered either to go to Rome or to transmit "their opinions carefully written." Which of these two courses of conduct did Copernicus adopt? The answer to this question is furnished by "that most distinguished man, Paul, bishop of Fossombrone, who was then in charge of this matter." [15] Paul of Middelburg (1445–1533), bishop of Fossombrone, in a published report to Leo X about the outcome of that pope's efforts to stimulate projected corrections of the defects in the current calendar, listed Copernicus among those who wrote, not among those who traveled to the Eternal City. [16] On this occasion, then, Copernicus did not go to Rome, nor was he in any special way "called to Rome."

While saying that "Copernicus was called to Rome," was Galileo perhaps thinking of Regiomontanus (1436–1476)? According to a popular historian's account, of which seven editions (four in Latin and three in Italian) were in circulation in Galileo's younger days, Regiomontanus "was made bishop of

---

world (Chicago: Encyclopaedia Britannica, 1952), XVI, 509. In mistranslating Copernicus' five simple Latin words, Wallis committed four blunders: he omitted "tum"; he mistook the tense of "praeerat"; and he misunderstood its meaning, as well as that of "negotio."

[9] Marzi (cited in n. 3, above), p. 78.

[10] Marzi, p. 79.

[11] *Loc. cit.*

[12] *Loc. cit.* (21 July 1514).

[13] Marzi, pp. 80–81 (24 July 1514).

[14] 1 June 1515 (Marzi, pp. 167–168); 8 July 1516 (Marzi, pp. 185–186).

[15] Drake (p. 178) transforms Galileo's "Vescovo Semproniense, allora soprintendente a quest' impresa" (NE, V, 312.11–12) into the "Bishop of Culm, then superintendent of this matter." Paul of Middelburg, bishop of Fossombrone, published several treatises on the calendar, whereas no such interest was shown by the bishop of Kulm. Drake's error is all the more surprising because he says (pp. vii–viii) that he based his translation upon the earlier English version "corrected and modernized." Yet according to the previous translator, Thomas Salusbury, *Mathematical collections and transla-*

tions, tome I (London, 1661), part I, p. 430, "the Bishop of *Sempronia*" was "at that time Super-intendent in that Affair." Dorothy Stimson, having failed to recognize that Copernicus' "Semproniensi" was merely a shortened form of "Foro Semproniensi," turned the bishop of Fossombrone into "a bishop from Rome" (*The gradual acceptance of the Copernican theory of the universe*; New York; also Hanover, New Hampshire; 1917, p. 115). Giorgio de Santillana mistakenly made "Cardinal Schönberg, then president of the Commission on the Calendar" instead of Paul of Middelburg (*The crime of Galileo*, University of Chicago Press, 1955, p. 22). For Paul of Middelburg, see Dirk Jan Struik, Paulus van Middelburg, *Mededeelingen van het nederlandsch historisch Instituut te Rome*, 1925, 5: 79–118; *idem*, Paolo di Middelburg e il suo posto nella storia delle scienze esatte, *Period. Mat.*, 1925, series 4, 5: 337–347; *idem*, Sull' opera matematica di Paolo di Middelburg, *R. C. Accad. Lincei*, 1925, series 6, 1: 305–308.

[16] Paul of Middelburg, *Secundum compendium correctionis calendarii* (Rome, 1516), fol. b1r.

Regensburg by Sixtus IV [17] and was called to Rome" [18] for the purpose of correcting the calendar. In like manner Leo X wrote to Paul of Middelburg on 16 February 1514 as follows:

> I have great need of your ability and erudition in computing and investigating chronological matters related to the Roman calendar as well as in the items on the agenda of the sacred Lateran Council. I therefore urge you to come to Rome at the very earliest time convenient to you, for your presence here is of importance to me.[19]

Whether confusion with Regiomontanus or Paul of Middelburg or somebody else be the explanation of Galileo's second error, he committed the third by saying about Copernicus that "Having reduced his system into six books, he published these at the instance of the Cardinal of Capua and the Bishop of Culm. And since he had assumed his laborious enterprise by order of the supreme pontiff, he dedicated this book *On the celestial revolutions* to Pope Paul III." [20] By injudiciously omitting the Italian words, "al suo successore, ciò è," Drake's new translation may give a false impression to the general reader, whom he has "principally in mind" (p. vii). Even students may be inclined to infer that, according to Galileo, Copernicus dedicated his *Revolutions* to the same supreme pontiff by whose order he had assumed his laborious enterprise. Actually Galileo says that the order emanated from a supreme pontiff, and "to his successor, that is, to Paul III," Copernicus dedicated the *Revolutions*.[21] In Galileo's time no Italian needed to be reminded that Paul III was the successor, twice removed, of Leo X. The latter was the supreme pontiff by whose order Copernicus had assumed his laborious enterprise. At any rate, that is what Galileo says (or implies) in the *Letter to the Grand Duchess*. But he makes no such statement in his preliminary formulation (the letter to Dini of 16 February 1615).[22] Like Galileo's description of Copernicus as a priest, his contention that Copernicus wrote the *Revolutions* by order of the pope emerges for the first time in the *Letter to the Grand Duchess*.

Let us try to trace the development of Galileo's fanciful notion about the origin of the *Revolutions*. Copernicus had said, as we saw above, that the only

---

[17] Actually Regiomontanus was not made a bishop by Pope Sixtus IV. The astronomer's alleged elevation to the episcopacy occurred only in the sympathetic imagination of this historian, Paolo Giovio (1483–1552), who was himself bishop of Nocera. In thus generously but gratuitously granting Regiomontanus a diocese, Giovio was operating in the realm of legend, not history; see Ernst Zinner, *Leben und Wirken des Johannes Müller von Königsberg, genannt Regiomontanus* (*Schriftenreihe zur bayerischen Landesgeschichte, 31*; Munich, 1938), p. 178. Yet in a chapter explicitly devoted to demolishing the legend of Regiomontanus, Lynn Thorndike repeated the legend that "he was made bishop of Regensburg" (*Science and thought in the fifteenth century*, New York: Columbia University Press, 1929, p. 146).

[18] Giovio, *Elogia veris clarorum virorum imaginibus apposita* (Venice, 1546), fol. 75r; *Elogia doctorum virorum*, ed. Antwerp, 1557, p. 271; ed. Basel, 1571, p. 287; *Elogia virorum literis illustrium*, Basel, 1577, p. 218: "creatus est a Xysto Quarto Ratisponensis Episcopus, accitusque Romam"; *Le iscrittioni poste sotto le vere imagini de gli huomini famosi*, tr. by Hippolito Orio (Florence, 1552), p. 228; edd. Venice, 1558, 1559, p. 263; cf. *An Italian portrait gallery*, tr. by Florence Alden Gragg (Boston: Chapman and Grimes, 1935), p. 163.

[19] Pietro Bembo, *Epistolarum Leonis decimi pontificis max. nomine scriptarum libri xvi* (Venice, 1535), book 7, no. 18; ed. Lyon, 1538, p. 157; ed. Basel, 1539, p. 272; ed. Lyon, 1540, pp. 166–167; ed. Basel, 1566, pp. 260–261; ed. Cologne, 1584, p. 167; ed. Strasbourg, 1611, pp. 147–148; in *Epistolarum familiarium libri vi*, ed. Venice, 1552, II, 204.

[20] Drake, p. 178; NE, V, 312.19–24: "avendo egli ridotta tal dottrina in sei libri, la pubblicò al mondo a i preghi del Cardinal Capuano e del Vescovo Culmense; e come quello che si era rimesso con tante fatiche a questa impresa d'ordine del Sommo Pontefice, al suo successore, ciò è a Paolo III, dedicò il suo libro delle Revoluzioni Celesti."

[21] No misunderstanding can possibly result from Salusbury's translation (cited in n. 15, above): Copernicus assumed "this so laborious an enterpize by the order of the Pope; he dedicated his book *De Revolutionibus Coelestibus* to His Successour, namely Paul III."

[22] Misdated "1614" by Guido Horn D'Arturo in his article on Copernicus in the *Enciclopedia italiana*, XI (1931), 318.

reason why the calendar was not reformed by the Fifth Lateran Council was that

> . . . the lengths of the year and month and the motions of the sun and moon had not yet been adequately measured. From that time on I have directed my attention to a closer study of these topics, at the instigation of that most distinguished man, Paul, bishop of Fossombrone, who was then in charge of this matter.

This instigation or admonition ("admonitus") by Paul of Middelburg becomes an order ("ordine") in Galileo's letter to Dini. But there the order is not yet a papal order, and it is still confined, as in Copernicus' own statement, "to the investigation of these periodic times." [23] In the *Letter to the Grand Duchess*, however, the task of ascertaining these times is given ("dato il carico") to Copernicus by Paul of Middelburg, whose power to issue orders is now transferred to the pope; and the papal order now embraces Copernicus' entire work in six books, not merely the portion dealing with the periodic times.

We have watched the actual admonition becoming improperly enlarged, first, into an "order," and then into a "papal order," whose subject matter expanded at the same time without any warrant from a part to the whole of the volume. But the bulk of the *Revolutions* was written long before the Fifth Lateran Council abandoned its unsuccessful effort at calendar reform; and it was this abandonment which induced Paul of Middelburg to admonish Copernicus to make "a closer study of these topics." In short, Galileo committed a grave blunder in saying that Copernicus "assumed his laborious enterprise by order of the supreme pontiff."

Not every work composed by Copernicus' contemporaries was the spontaneous creation of their own genius. For example, on the titlepage [24] of a plan for correcting the Roman calendar two Viennese astronomers prominently displayed the assertion that their joint effort had been written and published "at the request" of the pope and the Holy Roman Emperor; in the dedication these astronomers said that they wrote "by order" of the pope and emperor.[25] In like manner an Italian astronomer declared that he had computed his new ecclesiastical calendar "by order of popes Julius II, Leo X, Clement VII, and Paul III." [26] Copernicus said no such thing about his *Revolutions*.

That work is the subject of Galileo's fourth error, according to which, "When printed, the book was accepted by the holy Church, and it has been read and studied by everyone without the faintest hint of any objection ever being conceived against its doctrines." [27] Yet on 4 June 1539 in the home of Martin Luther (1483–1546), the initiator of German Protestantism, "mention was made of a certain new astronomer who proved that the earth moves, not the heavens, sun and moon,[28] just as anybody riding in a wagon or a boat would suppose that he is still and that the earth and trees are moving." Although

---

[23] NE, V, 293.17: "all' investigazione di tali periodi."

[24] Andreae Stiborii . . . et Georgii Tannstetter . . . super requisitione sanctissimi Leonis papae X et divi Maximiliani imperatoris . . . *De romani calendarii correctione* consilium . . . conscriptum et editum (Vienna, 1514).

[25] "Lucubrationes nostras quas . . . summi Pontificis et Caesareae Maiestatis iussi conscripsimus. . . ."

[26] Luca Gaurico (1475–1558), *Calendarium ecclesiasticum novum . . . iussu summorum pontificum Iulii II, Leonis X, Clementis VII et Pauli III . . .* supputatum (Venice, 1552).

For a later example of a calendaric work executed in obedience to a papal command, see n. 71, below.

[27] Drake, pp. 178–179; NE, V, 312.23–26.

[28] This misrepresentation of Copernicus as denying the moon's motion proves that Luther and his interlocutors lacked even the most rudimentary information about the astronomer whom they were discussing. But their conversation took place some four years before the publication of the *Revolutions* (for the date of the conversation, see *D. Martin Luthers Werke*, Weimar edition, *Tischreden*, IV, p. XIV, no. 4638).

the new astronomer was not mentioned by name, the reference was unmistakably to Copernicus, about whom Luther at once proceeded to say: "But that is how things go nowadays. Anyone who wants to be clever must not let himself like what others do. He must produce his own product, as this man [29] does, who wishes to turn the whole of astronomy upside down. But I believe in Holy Scripture, since Joshua ordered the sun, not the earth, to stand still." [30]

Luther's principal assistant, Philipp Melanchthon (1497–1560), on 16 October 1541 addressed a letter to the physician and mathematician Burkard Mithobius (Mithoff, 1501–1564) in which Melanchthon, the preceptor of Germany, as his admirers styled him, declared that "certain people believe it is a marvelous achievement to extol so crazy a thing, like that Polish astronomer who makes the earth move and the sun stand still. Really, wise governments ought to repress impudence of mind." [31] In a textbook first published in 1549 Melanchthon wrote:

> Out of love for novelty or in order to make a show of their cleverness, some people have argued that the earth moves. They maintain that neither the eighth sphere nor the sun moves, whereas they attribute motion to the other celestial spheres, and also place the earth among the heavenly bodies. Nor were these jokes invented recently. There is still extant Archimedes' book on *The sand-reckoner* in which he reports that Aristarchus of Samos propounded the paradox that the sun stands still and the earth revolves around the sun.[32]
>
> Even though subtle experts institute many investigations for the sake of exercising their ingenuity, nevertheless public proclamation of absurd opinions is indecent and sets a harmful example.

After citing some Biblical passages, Melanchthon continued:

> Encouraged by this divine evidence, let us cherish the truth and let us not permit ourselves to be alienated from it by the tricks of those who deem it an intellectual honor to introduce confusion into the arts.[33]

Melanchthon's son-in-law and editor, Kaspar Peucer (1525–1602), professor of mathematics at the university of Wittenberg, followed his father-in-law's advice by omitting from a primer all discussion of Copernicus "lest beginners be offended or disturbed by the novelty of his hypotheses." [34] Later,

[29] According to the first edition (Eisleben, 1566) of the *Tischreden*, Luther called Copernicus a fool ("Narr," fol. 580r; cf. Weimar ed., *Tischreden*, I, 419). An utterly unconvincing attempt to get rid of "Narr" as an interpolation was made by Wilhelm Norlind (Copernicus and Luther: a critical study, *Isis*, 1953, 44: 273–276), and decisively refuted by Heinrich Meyer (*Isis*, 1954, 45: 99).

[30] *Luthers Werke*, Weimar ed., *Tischreden*, IV (1916), pp. 412–413, no. 4638.

[31] *Alter libellus epistolarum Philippi Melanthonis* (Wittenberg, 1570; reprinted 1574), pp. 334–335. Since Galileo's time, Melanchthon's letter to Mithobius has been made more readily accessible in *Corpus reformatorum*, IV (Halle, 1837), 679.

[32] Thomas L. Heath, *The works of Archimedes* (Cambridge, England, 1897; reprinted, New York: Dover Publications, 1953), pp. 221–222.

[33] Melanchthon, *Initia doctrinae physicae* (Wittenberg, 1549), fol. 47v–48v, reprinted in *Corpus reformatorum*, XIII (Halle, 1846), 216–217. In the second (1550) and subsequent editions of his *Initia* Melanchthon modified his condemnation of Copernicus; see Wohlwill, *Melanchthon und Copernicus*, *Mitt. Gesch. Med.*

Naturw., 1904, 3: 260–276. Like Melanchthon, Jean-Pierre de Mesmes, *Les institutions astronomiques* (Paris, 1557), p. 56, bk. I, ch. 19, linked Copernicus with Aristarchus, and rejected their "absurd opinion" and "false hypothesis or proposition." A pupil of Melanchthon, Michael Stanhuf, *De meteoris* (Wittenberg, 1562; reprinted, 1578), fol. C6v–7r, labeled the belief in the earth's motion a "silly and absurd opinion," held by "utterly crazy and insane" people such as Aristarchus and certain unnamed modern writers.

[34] Peucer, *Elementa doctrinae de circulis coelestibus et primo motu* (Wittenberg, 1551), fol. G1v; Peucer's *Elementa* with its dread of Copernicanism was reprinted at Wittenberg in 1553, 1558, 1563, 1569, 1576 and 1587. Copernicus' name was not mentioned by Cornelius Valerius, a highly influential professor who declared in his textbook on natural philosophy: "There have been those who thought that the heavens were motionless while the earth moved; their false opinion needs no special refutation" (*Physicae . . . institutio*, ed. Lyon, 1568, p. 18; edd. Antwerp, 1574, 1575, p. 17; ed. Marburg, 1591, p. 30; ed. Antwerp, 1593, pp. 15–16; ed. Marburg, 1593, p. 32). In like manner Copernicus, although not named, was compared

in a more advanced work, Peucer denounced Copernicanism as absurd, far from the truth, offensive and not fit to be taught in the schools.[35]

An imaginary dialogue between a schoolmaster and his scholar contained the earliest discussion of Copernicanism in an English book. Its author, Robert Recorde (c. 1510–1558), had the scholar describe Copernicus' essential ideas as "such vaine phantasies, so farre againste common reason, and repugnante to the consente of all the learned multitude of Wryters." [36] This attitude toward Copernicanism was expressed in the work which remained the standard introduction to astronomy in England throughout the latter half of the sixteenth century. Shortly before its second edition appeared in 1596, a successful English popularizer of science, Thomas Blundeville (fl. 1560–1602), referred to Copernicus' "false supposition." [37]

The eminent scholar Julius Caesar Scaliger (1484–1558), "that violent and passionate man," [38] put the name of Copernicus in the margin as a sidenote alongside the recommendation that certain "writings should be expunged or their authors whipped." [39] With equal severity the Sicilian mathematician Francesco Maurolico (1494–1575) said: "Nicholas Copernicus, who maintained that the sun is still and the earth has a circular motion, deserves a whip or a scourge rather than a refutation." [40]

In a didactic poem on astronomy the Scottish humanist and historian George Buchanan (1506–1582) unquestionably aimed the following verses at Copernicus:

> Buried in opaque darkness, ignorance has not yet ceased to bark out loud, rashly condemning the heavens to motionlessness and whirling the sluggish earth in a swift motion.[41]

Jean Bodin (1530–1596), the French philosopher, charged Copernicus with supposing "two absurd things." One of these alleged absurdities was that "the earth undergoes the movements which all the astronomers have always assigned to the heavens. . . . It is even more strange to put the sun in the center of the universe, and the earth fifty thousand leagues away from the center." Bodin argued further that "since the earth is one of the simple bodies, like the heaven and the four elements, we must conclude that it can have only a

---

with "phrenetic spirits" by the French poet Du Bartas (1544–1590) in *La Sepmaine* (Paris, 1578), fourth day, lines 121–164. More than three dozen editions of this extremely popular cosmographical poem were published before Galileo wrote the *Letter to the Grand Duchess*; see *The Works of Guillaume de Salluste, Sieur du Bartas* (Chapel Hill, 1935–1940), I, 70–77.

[35] Peucer, *Hypotheses astronomicae* (Wittenberg, 1571), Dedication, fol.)(3r, 5v. In similar fashion Thomas Hill (fl. 1553–1575), author of a posthumously published textbook on astronomy, *The Schoole of Skil* (London, 1599), decided that Copernicus' reasoning would "offend and trouble the young students in the Art" (quoted by E. G. R. Taylor, *The Mathematical Practitioners of Tudor and Stuart England*, Cambridge, 1954, p. 336).

[36] Recorde, *The Castle of Knowledge* (London, 1556), p. 165.

[37] Blundeville, *Exercises* (London, 1594), fol. 181r; 4th ed., London, 1613, p. 380; later edd. were published after Galileo's *Letter to the Grand Duchess*. According to Thomas S. Kuhn, *The Copernican Revolution* (Cambridge: Harvard University Press, 1957), p. 186, Blundeville made this remark "in the preface of an astron-

omy text." Actually it appears in Book 2, Chapter 5, of "A plaine treatise of the first principles of cosmographie, and specially of the spheare," which has no preface.

[38] Vernon Hall, Jr., Life of Julius Caesar Scaliger, *Trans. Amer. phil. Soc.*, 1950, 40 (part 2): 85.

[39] J. C. Scaliger, *Exotericarum exercitationum liber quintus decimus*, Exercitatio XCIX, part 2 (Paris, 1557, fol. 142v).

[40] Maurolico, *Opuscula mathematica* (Venice, 1575), p. 26. A mistaken attempt to alter the traditional understanding of this passage is refuted by Edward Rosen, Maurolico's attitude toward Copernicus, *Proc. Amer. phil. Soc.*, 1957, 101: 177–194.

[41] Buchanan, *De sphera*, book II, lines 143–146 (in Buchanan's *Franciscanus et fratres*, Heidelberg, 1609, p. 220). Six other editions of Buchanan's *De sphera* were printed before Galileo wrote the *Letter to the Grand Duchess*. Buchanan's reaction to Copernicus was discussed by James R. Naiden, *The Sphera of George Buchanan*, 1952 (procurable from W. H. Allen, 2031 Walnut Street, Philadelphia 3, Pennsylvania), pp. 52–54.

single motion which is proper to it. Yet Copernicus assigns it three different motions, of which it can have only one proper to it. The others would be violent, but this is impossible." [42] In a subsequent treatise, written while "all of France was aflame in civil war," [43] Bodin declared that the belief in a motionless sun and moving earth "was revived in our time by Copernicus, but it can easily be refuted by its own vacuity. . . . Copernicus' opinion gives rise to very grave absurdities." If he is right about the earth's motion, "all the foundations of physics must crumble. . . . No one who is in his right mind or who has had the slightest training in the physical sciences will ever believe that the dense and solid earth with its heaviness and weight simultaneously moves up and down, about its own center, and around the sun, while performing a libration." [44]

Tycho Brahe (1546–1601), the great Danish astronomer, asked:

> What need is there without any justification to imagine the earth, a dark, dense and inert mass, to be a heavenly body undergoing even more numerous revolutions than the others,[45] that is to say, subject to a triple motion, in violation not only of all physical truth but also of the authority of Holy Scripture, which ought to be paramount? [46]

According to Brahe, when Copernicus

> . . . stated that the earth's dense and inert mass, which is unsuitable for movement, is active in a course of motion (indeed, a threefold course) no less regular than the luminaries of the aether, he opposed . . . not only the principles of physics but also the authority of Holy Scripture, which several times confirms the immobility of the earth.[47]

Brahe maintained that

> . . . the earth, which we inhabit, occupies the center of the universe and does not perform any annual motion such as Copernicus supposed. These propositions must be upheld without any doubt; so I believe, together with the ancient astronomers and the accepted opinions of the physicists, supported by Holy Scripture.[48]

The earth's annual revolution "does not occur at all," such ideas being "not only dubious but obviously false and absurd." [49] Brahe insisted on the "absurdity of this Copernican arrangement of the revolutions in the universe." [50] Copernicus' "arrangement of the apparent orbits in the bodies of the universe does not in fact correspond with the truth." [51] From Copernicus' ascription to the earth of an annual revolution around a motionless sun "some absurdity

[42] Bodin, *Les six livres de la republique*, book 4, chapter 2 (Paris, 1576, p. 442); of this work there were, before Galileo wrote his *Letter to the Grand Duchess*, some eleven editions in the original French, besides versions in Italian, Spanish and English. When Bodin translated his *Republic* into Latin (Lyon, 1586, followed by four other editions), he modified the wording of his anti-Copernicus passage, but not the reasoning.
[43] Bodin, *Universae naturae theatrum* (Lyon, 1596; Frankfurt, 1597; Hanau, 1605), p. 633.
[44] *Op. cit.*, pp. 580–582. In the French translation by François de Fougerolles (Lyon, 1597) the quoted passages occur at pp. 837–840.
[45] Lynn Thorndike, *A history of magic and experimental science* (New York: Columbia University Press, 1923–1958), VI, 52, paraphrased Tycho's question as follows: "Why make our . . . earth a star and more revolved than the others?" But "multiplicius quam caetera revolutum" means "undergoing more revolutions

than the others," not "more revolved than the others," if indeed this expression of Thorndike's may be said to have any meaning at all.
[46] *Astronomiae instauratae progymnasmata* (Prague, 1602; Frankfurt, 1610), p. 661; more readily accessible, since Galileo's time, in *Tychonis Brahe dani opera omnia*, ed. J. L. E. Dreyer (Copenhagen, 1913–29), III, 175.13–17.
[47] Brahe, *De mundi aetherei recentioribus phaenomenis* (Uraniborg, 1588; Prague, 1603; Frankfurt, 1610), pp. 186–187; more readily accessible, since Galileo's time, in *Tychonis opera*, ed. Dreyer, IV, 156.16–21.
[48] *Op. cit.*, edd. 1588, 1603, 1610, p. 187; *Opera*, IV, 156.34–37.
[49] *Astronomiae instauratae progymnasmata*, edd. 1602, 1610, p. 548; *Opera*, III, 63.9–11.
[50] *Op. cit.*, edd. 1602, 1610, p. 549; *Opera*, III, 63.40–41.
[51] *Op. cit.*, edd. 1602, 1610, p. 11 ; *Opera*, II, 14.24–25.

arises, only for physicists, but not for mathematicians." [52] "By ordaining a triple motion of the earth, Copernicus introduced no trivial physical absurdities." [53] Brahe emphatically denied that the "physical absurdities which accompany the Copernican hypothesis were adequately refuted by him." [54]

A letter in which Brahe, the foremost astronomer of the second half of the sixteenth century, referred to the "absurdities introduced by Copernicus" was published by the recipient, Giovanni Antonio Magini (1555–1617). [55] Long before receiving Brahe's letter, Magini himself had publicly referred to "absurd hypotheses, such as Copernicus imagined." [56] "It seems to Copernicus, against all truth and philosophy, that the earth moves, while the sun and the eighth sphere are motionless. . . . Copernicus' opinion about the motion of the earth is erroneous." [57] His hypotheses "are attacked by nearly everybody for being too far away from the truth and absurd." [58] These condemnations of Copernicus came from Magini's pen shortly after he was appointed to fill the vacant professorship of mathematics at the university of Bologna. An unsuccessful rival for the same appointment had been none other than Galileo himself. [59]

Between Maurolico's attack on Copernicus in 1575 and Magini's in 1589, a third Italian scientist, Francesco Barozzi, in 1585 declared that Copernicus "followed the false opinion of Aristarchus"; [60] when Barozzi referred to the idea of the earth's motion, he labeled it in a sidenote "the false opinion of Aristarchus and Copernicus." [61]

Among Galileo's contemporaries in Italy the most renowned for his knowledge of astronomy was Christopher Clavius (1538–1612), to whom Galileo, like Maurolico, turned for help. In the first edition of a work which passed through half a dozen editions (plus a dozen re-impressions) Clavius said that Copernicus' "idea conflicts with many aspects of experience and the common opinion of all philosophers and astronomers." [62] In the second edition Clavius inserted the additional condemnation that "many absurdities and errors are contained in Copernicus' position." [63] Finally, in the fourth edition, Clavius supplemented his previous criticisms by asserting that Copernicus

> . . . assumes hypotheses which are quite unsound, absurd and out of line with the common sense of mankind, not to say foolish, when he deprives the sun of all motion and stations it in the center of the universe, but endows the earth with a multiple motion and places it, together with the other elements and the sphere of the moon in the third heaven, between Venus and Mars. [64]

This opinion of Copernicus that the earth moves was declared by Giulio Cesare LaGalla (1571–1624), professor of logic at the university of Rome,

---

[52] *Apologetica responsio ad Craigum scotum de cometis* (Uraniborg, 1591); *Opera*, IV, 446.22–23. Similar mathematical approval of Copernicus had been coupled with physical disapproval in Brahe's *De disciplinis mathematicis oratio*, a lecture delivered at the University of Copenhagen in 1574 and published in 1610 (*Opera*, I, 149.30–33).

[53] *Opera*, IV, 473.38–39.

[54] *Epistolarum astronomicarum libri* (Uraniborg, 1596; Nuremberg, 1601; Frankfurt, 1610), p. 147; *Opera*, VI, 177.7–8.

[55] Magini, *Tabulae primi mobilis* (Venice, 1604), fol. 8ov; reprinted in Brahe's *Opera*, ed. Dreyer, VII, 293.22, and in Antonio Favaro, *Carteggio inedito di T. Brahe . . . con G. A. Magini* (Bologna, 1886), p. 399.

[56] Magini, *Novae coelestium orbium theoricae* (Venice, 1589), address to the reader, fol. b2v; ed. Mainz, 1608, fol. B1v.

[57] *Op. cit.*, ed. 1589, fol. b4v; ed. 1608, fol. B4r.

[58] *Op. cit.*, preface, ed. 1589, fol. c5v; ed. 1608, fol. C1r.

[59] Favaro, Galileo ed il Magini aspiranti ad una lettura di matematica nello Studio di Bologna, *Atti Ist. veneto*, 1922–1923, *82* (part 2): 148–155.

[60] Barozzi, *Cosmographia* (Venice, 1585), preface, fol. b4r.

[61] *Op. cit.*, p. 35.

[62] Clavius, *In sphaeram Ioannis de Sacro Bosco commentarius* (Rome, 1570), p. 87.

[63] *Op. cit.*, 2d ed. (Rome, 1581), p. 437.

[64] *Op. cit.*, 4th ed. (Lyon, 1593), p. 68; for a later condemnation of Copernicus by Clavius, see n. 77, below.

to be "obviously absurd, in opposition to and in conflict with the common sense of all men, educated and uneducated." [65] LaGalla was convinced that he had shown "Copernicus' opinion to be false and impossible." [66]

The foregoing array of published pronouncements by such leading luminaries of the sixteenth century as the religious reformers Luther, Melanchthon and Peucer, the critic J. C. Scaliger, the poet Buchanan, the philosopher Bodin, the mathematicians Maurolico and Barozzi, the astronomers Brahe, Magini and Clavius, and the peripatetic LaGalla, shows how utterly mistaken was Galileo's statement that Copernicus' *Revolutions* "has been read and studied by everyone without the faintest hint of any objection ever being conceived against its doctrines." [67]

The fifth and last error of Galileo with which we shall be concerned occurs in the assertion: "Since that time not only has the calendar been regulated by his [Copernicus'] teachings, but tables of all the motions of the planets have been calculated as well." [68] In referring to the regulation of the calendar, Galileo had in mind the reform of the calendar promulgated in antiquity by Julius Caesar. This so-called Julian calendar had been transformed into the Gregorian calendar by order of Pope Gregory XIII in 1582. Was the Gregorian calendar regulated "in conformity with Copernicus' doctrine" ("conforme alla sua dottrina"), as Galileo claimed?

In the bull announcing the new calendar Gregory XIII said:

> Antonio Giglio, doctor of arts and medicine . . . brought me a book written some time ago by his brother Aloisio. In this book Aloisio shows that by means of a certain new cycle of epacts devised by him . . . all the defects in the calendar can be remedied in accordance with a fixed rule that will endure throughout all the ages so that the calendar apparently will never require any change hereafter.[69]

Hence the Gregorian calendar was regulated in conformity with the doctrine of Aloisio Giglio, not Nicholas Copernicus.

Giglio's "new method of restoring the calendar," the pope went on to say, "is contained in the thin volume which I sent a few years ago to the Catholic rulers and more famous universities." [70] This thin volume or *Compendium* [71]

---

[65] LaGalla, *De phoenomenis in orbe lunae* (Venice, 1612), p. 14; reprinted in NE, III, 337.18–20. LaGalla and his book were discussed by Edward Rosen, *The naming of the telescope* (New York, 1947), pp. 54–59. In NE, XX, 465, Favaro gave the year of LaGalla's birth as 1576, and the day of his death as 15 March 1624. Perhaps he took these dates from Gerolamo Boccardo, *Nuova enciclopedia italiana*, 6th ed., XII (Turin, 1881), 71, which cites the biography of LaGalla by his pupil Leone Allacci (*Iulii Caesaris LaGallae philosophi romani vita*, Paris, 1644). But Allacci says (p. 1) that LaGalla was born in 1571, not 1576. The earlier date is no misprint, for Allacci states that LaGalla wrote his *De immortalitate animorum* (Rome, 1621) when he was fifty years old ("annum agens quinquagesimum," p. 9). Moreover, LaGalla composed an epitaph for himself in 1623, the fifty-second year of his life ("anno suae aetatis LII; MDCXXIII," Allacci, p. 23), and his death in 1624 occurred in his fifty-third year ("eius vitae fuerat quinquagesimus tertius," Allacci, p. 22). The latter records the day of LaGalla's demise as "16. Cal. Martii," which is the fifteenth day of February, not March.

[66] LaGalla, p. 21; NE, III, 347.28. Galileo's handwritten comments in his copy of LaGalla's book were printed in NE, III, at the foot of pp. 323–387 passim, and on pp. 393–399. Galileo wrote to a friend about LaGalla's book prior to 17 March 1612, more than three years before he finished his *Letter to the Grand Duchess* (NE, XI, 283.55).

[67] Galileo's assertion in the *Letter to the Grand Duchess* that there were no objections to Copernicanism may be regarded as a negative and diluted version of the statement in his preliminary formulation that Copernicus' "doctrine was later followed by everybody" (NE, V, 293.20). Both versions are equally unhistorical.

[68] Drake, p. 178; NE, V, 312.17–19.

[69] *Bullarum diplomatum et privilegiorum s. romanorum pontificum taurinensis editio* (Turin, 1857–72), VIII, 386–387.

[70] *Op. cit.*, VIII, 387.

[71] *Compendium novae rationis restituendi calendarium*, reproduced in Clavius, *Romani calendarii a Gregorio XIII. P.M. restituti explicatio* (Rome, 1603), pp. 3–12, and in Clavius, *Opera mathematica* (Mainz, 1611–12), V, 3–12. Clavius' *Explicatio*, which was "published by order of Pope Clement VIII," may be grouped with the command performances discussed in nn. 24–26, above.

pointed out that Giglio accepted the length of the year given by the Alfonsine Tables. According to these Tables, which had been sponsored by King Alfonso X of Castile in the thirteenth century, the length of the (tropical) year was constant. But Copernicus knew that this length varied.[72] Giglio decided against Copernicus and in favor of the medieval Tables, on the ground that the Alfonsine length "is an average of the various measurements and therefore less subject to error."[73] In this basic matter of the length of the year, then, Giglio, the principal author of the reform, maintained that the calendar should not be regulated in conformity with Copernicus' doctrine.

The writers of the *Compendium* added that

> . . . if anyone thinks the Alfonsine calculations too uncertain to be trusted and prefers adhering to more recent authorities, he will surely understand that this ingenious cycle and table of epacts devised by Giglio are so arranged and disposed that they can be adapted without any trouble to the calculations of Copernicus or anybody else if the set of equations recently prepared is substituted for the one which we wrote in the margin.[74]

Like Giglio and the *Compendium*, the Gregorian calendar decided against Copernicus. Adopting the Alfonsine length of the year, it promulgated a rule requiring the omission of three leap days in four centuries. Clavius, a member of the papal commission which recommended the reform of 1582,[75] was delegated to defend the new calendar against its critics. With regard to the rejection of Copernicus, Clavius explained that "in celebrating Easter, the church ought to follow something . . . not far from the truth rather than the precise calculation of the astronomers."[76] After all, the task confronting the church in undertaking to reform the calendar was not so much the solution of a theoretical scientific question as the elimination of a pressing practical problem: the time was out of joint. And of all the astronomers, surely the last to be followed was Copernicus, whose hypotheses, said Clavius, were "uncertain, not to say absurd, conflicting with the common opinion of mankind, and rejected by all students of nature."[77] Clavius agreed, then, with Giglio and the *Compendium* that the calendar should not be regulated in conformity with Copernicus' doctrine. And in fact, despite Galileo's misstatement, the Gregorian calendar was not regulated in conformity with Copernicus' doctrine.

We have now examined one by one Galileo's five misstatements about Copernicus in the *Letter to the Grand Duchess*. Let us put them side by side to see whether they have any feature in common. According to Galileo, (1) Copernicus was a priest; (2) he was called to Rome; (3) he wrote the

[72] Michael Maestlin, *Alterum examen novi pontificialis gregoriani kalendarii* (Tübingen, 1586), p. 5: "The Prussian Tables make a distinction between the true tropical year and the mean tropical year. They maintain that the true tropical year is sometimes longer and sometimes shorter . . . as Copernicus exhaustively proves. This variation is absolutely unknown to the Alfonsine Tables." One of the chapters in Copernicus' *Commentariolus* is entitled "Equal Motion Should Be Measured Not by the Equinoxes but by the Fixed Stars"; see Edward Rosen, *Three Copernican treatises* (New York, 1939), p. 65; at pp. 114–117 and 127–131 will be found a summary of Copernicus' position by his disciple George Joachim Rheticus. Wallis' translation of the *Revolutions* (cited in n. 8, above), XVI, 622–674, may serve to remind us how wise Roger Bacon was in

insisting that a translator should understand not only the language but also the content of the document he is trying to translate.

[73] *Compendium*, in Clavius, *Explicatio*, p. 4, or *Opera*, V, 3.

[74] *Compendium*, in Clavius, *Explicatio*, p. 11, or *Opera*, V, 11.

[75] Ferdinand Kaltenbrunner, *Beiträge zur Geschichte der Gregorianischen Kalenderreform*, *Sitzungsberichte der k. Akademie der Wissenschaften*, phil.-hist. Classe. Vienna, 1881, 97: 54.

[76] Clavius, *Novi calendarii romani apologia* (Rome, 1588), p. 38; p. 20 of the reprint in Clavius, *Opera*, V.

[77] *Op. cit.*, ed. 1588, p. 29; ed. *Opera*, p. 16. This condemnation of Copernicus by Clavius should be added to the passages cited in nn. 62–64, above.

Revolutions by order of the pope; (4) his book was never adversely criticized; (5) it was the basis of the Gregorian calendar. Actually, Copernicus was not a priest; he was not called to Rome; [78] he did not write the *Revolutions* by order of the pope; the book received much adverse criticism, particularly on the ground that it contradicted the Bible; it was not the basis of the Gregorian calendar. If we compare Galileo's five misstatements with the truth, we see that each of them tended to bind Copernicus more closely to the Roman Catholic church.

Galileo made these five misstatements at a time when he was fighting hard to prevent his church from denouncing Copernicanism as heretical. This far-seeing and loyal purpose dominates his entire *Letter to the Grand Duchess*, an eloquent (albeit unavailing) effort to save the Roman Catholic church from committing a grievous error. For, by placing Copernican treatises on the Index of Prohibited Books in Galileo's lifetime, the Roman Catholic church made a mistake, as it implicitly acknowledged when it subsequently removed those same treatises from the Index. [79]

It was not any deliberate desire to distort the facts, but rather the intensity of his struggle against bigoted and narrow-minded coreligionists that, in my opinion, led Galileo astray into these five misstatements. [80] In only one instance (LaGalla's criticism of Copernicus) have we seen any evidence that Galileo should have been conscious of making a misstatement.

Consider, for example, Copernicus' dedication of his *Revolutions* to the pope. This unilateral action by Copernicus was interpreted by Tommaso Campanella (1568–1639), the Dominican defender of Galileo, to mean that "Pope Paul III . . . to whom Copernicus dedicated the book . . . approved it" [81] and gave his "permission that the book should be printed." [82] Galileo and Campanella both knew that it was customary for Italian authors to seek prior permission for a dedication. That Copernicus did likewise was an (unhistorical) assumption made by Campanella and perhaps by Galileo too. Similar considerations probably underlie his other misstatements, which should be considered honest mistakes rather than conscious falsehoods.

---

[78] Galileo's misstatement that Copernicus was called to Rome for the purpose of reforming the calendar was later transformed into the equally mistaken assertion that he was called to Rome to teach mathematics there. Of the numerous repetitions of this error, only one need be cited here: Belisario Ruiz Wilches, "La obra de Nicolas Copernico," in *Nicolas Copernico* (Bogota, 1943), p. 15. For a demolition of the legend that Copernicus was a professor at the University of Rome, see Ryszard Gansiniec, "Rzymska profesura Kopernika," in *Kwartalnik historii nauki i techniki,* 1957, *2*: 471–484, with a summary in English at pp. 482–484.

[79] In the opinion of Antonio Favaro (1847–1922), unquestionably the greatest Galileo scholar of all time, the condemnation of Galileo is "if not the greatest, one of the greatest errors of the Roman Curia" (*Galileo e l'inquisizione,* Florence, 1907, p. 9).

[80] The reason for his misstatement of an astronomical constant was analyzed by Edward Rosen, Galileo on the distance between the earth and the moon, *Isis,* 1952, *43*: 344–348.

[81] Campanella, *Apologia pro Galileo* (Frankfurt, 1622), p. 9; English tr. by Grant McColley, *Smith College studies in history,* 1936–1937, *22*: 10; for the quality of McColley's translation, see Edward Rosen, *Journal of the History of Ideas,* 1957, *18*: 440–443.

[82] Campanella, p. 54; McColley, p. 71. In this context Campanella's expression "a tot theologis approbata" was (mis)translated by McColley as "the approval of all theologians." Here the familiar Latin adjective "tot" ("so many") was evidently confused by McColley with the French word "tout" ("all"). What was approved by the theologians in question? According to Campanella, "the opinion of Copernicus and Galileo was approved by so many theologians." According to McColley, Campanella's reply to certain arguments "has the approval of all theologians." Yet a few lines further on, McColley himself had Campanella say that Copernicus was "supported by the authority of concurring theologians." Obviously McColley has not "even comprehended the text he is editing," as an eminent reviewer has justly said (*Journal of Philosophy,* 1939, *36*: 157).

21

Was Copernicus' Revolutions Annotated by Tycho Brahe

NICHOLAS COPERNICUS (1473-1543), the greatest astronomer of the first half of the sixteenth century, published his epoch-making treatise *On the Revolutions of the Heavenly Spheres* (*De revolutionibus orbium coelestium*)[1] in Nuremberg in 1543. The second edition (Basel, 1566) added, after Copernicus's *Revolutions,* at folios 196v-213r the *Narratio prima*[2] by George Joachim Rheticus (1514-1574),[3] Copernicus's only disciple. This Basel 1566 edition of Copernicus-plus-Rheticus was cited by Tycho Brahe (1546-1601), the greatest astronomer of the second half of the sixteenth century, in his *Preliminary Exercises for the Restoration of Astronomy* (*Astronomiae instauratae progymnasmata*): "The very learned astronomer George Joachim Rheticus, Copernicus's disciple, in that *Narratio,* [addressed] to Johannes Schöner, which is added at the end of Copernicus's work, says that the ancients

1. English translation and commentary by Edward Rosen (Baltimore: Johns Hopkins Univ. Press, 1978).

2. English translation and commentary by Edward Rosen, *Three Copernican Treatises,* 3rd rev. ed. (New York: Octagon, 1971), pp.107-96. Hereafter cited as "TCT."

3. Rheticus did not die in 1576, as in Cimelia, p.15/Description/9 (cited in n.12, below). Karl Heinz Burmeister, *G. J. Rhetikus* (Wiesbaden: Pressler, 1967-68), I, 176-77.

did not know the precise length of the sidereal year also for the reason that, having no sure theory of eclipses, they neglected to take account of the solar parallaxes, which are ultimately derived therefrom."[4]

Earlier in the same paragraph Brahe mentioned the mission on which he had sent a student to check the latitude of Frombork, the scene of most of Copernicus's observations. That expedition took place "in the course of the year 1584, four years ago."[5] Hence Brahe cited Copernicus-plus-Rheticus in 1588, in *Preliminary Exercises,* Tome I, which was first published posthumously in 1602. "This work was written between 1582 and 1592,"[6] according to an appendix which looks as though it had been composed by Tycho's heirs. Actually, it was drafted in that style by Tycho's foremost collaborator, Johannes Kepler (1571-1630), who privately informed an Italian astronomer on 1 February 1610: "I myself am the author of the Appendix to the *Preliminary Exercises,*"[7] Tome I. The printing of this work had begun while Tycho was still alive, but in 1601 it was not finished at the time of his unexpected death on 24 October. "From September on, I investigated the sun's eccentricity, the problem on which Tycho [was working when he] died," Kepler reports. "We spent a month taking care of him while he was sick, and attending to his burial after his death. Then, until Christmas [1601], I reread the *Preliminary Exercises,* compiled the index, and drafted the notes."[8]

Four years after Brahe cited Copernicus-plus-Rheticus, he belatedly found out that Rheticus had bequeathed his library to Brahe's friend Thaddeus Háyek (1525-1600), to whom Brahe wrote on 22 March 1592: "From a letter recently sent to me by Dr. [Johannes] Sager of Lübeck I learned that Rheticus's library was bequeathed to you in his testament. If so, I ask you to have whatever concerns astronomy, whether he ac-

4. *Tychonis Brahe Dani opera omnia* (Copenhagen: Gyldendal, 1913-29), II, 31/22-28. Hereafter cited as "тв." тст, p.117/6-7.

5. тв II, 30/7-8.

6. тв III, 320/7-8.

7. *Carteggio inedito di Ticone Brahe, Giovanni Keplero...con Giovanni Antonio Magini,* ed. Antonio Favaro (Bologna, Zanichelli, 1886), p.331/7↑. Johannes Kepler, *Gesammelte Werke* (Munich: Beck, 1937—), XVI, 279/27. Hereafter cited as "GW."

8. *Joannis Kepleri astronomi opera omnia,* ed. Christian Frisch (Frankfurt am Main and Erlangen: Heyder and Zimmer, 1858-71; rpt. I-II, Hildesheim: Gerstenberg, 1971-77), VIII, 742/8-11. Hereafter cited as "F." тв II, 440/n. 1/1-3.

quired it from his teacher Copernicus or by his own activity, copied at my expense and forwarded to me promptly."[9] The surviving Brahe-Háyek correspondence does not indicate whether Rheticus's astronomical legacy was copied for Brahe. In any case, Copernicus-plus Rheticus would probably not have been copied since that work, printed in 1566, was presumably still available in Basel and, besides, Brahe had his own copy which, as we just saw, he had cited in 1588.

After Tycho Brahe's death on 24 October 1601, his library passed into the hands of his younger son, George Brahe (1583-1640). On 16 November 1624 Kepler sent George a recommendation that "a careful catalog should be compiled of everything in the way of books and manuscripts still surviving from Tycho's library."[10] Kepler's advice was not heeded, and no such Tycho catalog was ever prepared.

Two years after the death of George Brahe, in 1642 the Jesuit College in Prague was given a part of his father's library, including Copernicus-plus-Rheticus.[11] On its title page is written: "Collegii Caesarei Societatis Jesu Pragae Anno 1642" (Property of the Imperial College of the Society of Jesus in Prague, in the year 1642). The same hand added: "Ex Bibliotheca et Recognitione Tichoniana"[12] (From the Library and Scrutiny of Tycho). On the flyleaf preceding the title page a different hand pointed out: "NB Insunt notae marginales manu Tychonis Brahe propria inscriptae." (Observe: there are present marginal notes written by Tycho Brahe's own hand.) After the suppression of the Jesuits in 1773 by Pope Clement XIV, together with the other Jesuit holdings in Prague, Copernicus-plus-Rheticus was incorporated in the library of

9. TB VII, 333/39-334/1.

10. GW XIX, 203/7-8. Kepler's recommendation is incompatible with the assumption in Wilhelm Prandtl, "Die Bibliothek des Tycho Brahe," *Philobiblon, Zeitschrift für Bücherliebhaber,* 5 (1932), 327/19↑-16↑ (rpt. Vienna: Reichner, 1933, p.16/5-8) that "Apparently Tycho's heirs let Kepler use Tycho's professional library."

11. The Jesuits did not receive this gift "right after the death of the great astronomer" in 1601, as in Richard Kukula, "Die Tychoniana der Prager K. K. Universitäts-Bibliothek," *Zeitschrift für Bücherfreunde,* 10 (1906-07), 19/10↑-8↑. The gift was mentioned by Josef Volf, "Tycho Brahe a jeho hvězdářska pozůstalost," *Český Bibliofil,* 3 (1931), p.18/6-8, not p.19, as in Kleinschnitzová (cited in n.21, below), p.76, n.3.

12. *Editio cimelia Bohemica,* xvi, ed. Zdeněk Horský (Prague: Bibliotheca Rei publicae socialisticae Bohemicae Pragensis, 1971). Hereafter cited as "Cimelia." These two entries on Cimelia's title page provide no support for its assertion (p.12/24-25) that "A librarian in the Prague Jesuit College Library wrote here in 1642 that the book was listed in the catalog." No catalog is mentioned.

Prague University (shelf mark 14 B 16—Tres M 11).[13] In 1971 a facsimile was issued as *Nicolai Copernici De revolutionibus orbium coelestium libri sex* (*editio Basileensis*) *cum commentariis manu scriptis Tychonis Brahe* [Nicholas Copernicus's *On the Revolutions of the Heavenly Spheres,* in six Books (Basel edition), with annotations written by the hand of Tycho Brahe].

This 1971 Cimelia facsimile declared: "Absolutely no one of the relatively numerous experts in historical science and bibliology who have written about this book has raised the slightest objection to the authenticity of Brahe's commentary."[14] Yet more than a century earlier, the distinguished biographer of Tycho Brahe and editor of his correspondence and cometary observations, Frederik Reinholdt Friis (1836-1910), said with regard to this Prague Copernicus-plus-Rheticus: "This copy is completely provided on nearly every page with added notes and corrections, but these could not be additions by the astronomer [Tycho Brahe], as has been assumed heretofore."[15]

This assumption, accepting the 1642 Jesuit description, had been made by the historian of the Prague University library, Joseph Adolph Hanslik (1785-1859).[16] Friis's rejection of the Jesuits' claim, endorsed by Hanslik, was overlooked by František Josef Studnička (1836-1903) who, in commemorating the tercentenary of Tycho's death, wrote: "A Copernicus annotated by a Brahe will not be found again."[17] This prediction was quoted with approval by Bořivoj Prusík (1872-1928), who remarked that from the annotations "one sees what position Brahe took with regard to several views of Copernicus. . . . How closely Brahe studied Copernicus's work and wanted to study it further is shown by the large

13. Copernicus-plus-Rheticus has not "been a permanent part of the Prague University Library since 1642," despite Cimelia, p.15/Description/3-4. Emma Urbánková, *Rukopisy a vzacné tisky pražské Universitní knihovny* (Prague: Statní pedagogické nakladatelství, 1957), p.72/8-11. Melba Berry Bennett, *The Story of Prague's Old Libraries* (Palm Springs: Welwood Murray Memorial Library, 1964), p.2.

14. Cimelia, p.12/14↑-12↑.

15. F. R. Friis, "Tyge Brahe's Haandskrifter i Wien og Prag," *Danske Samlinger,* 4 (1868-69), 267/7-10.

16. J. A. Hanslik, *Geschichte und Beschreibung der Prager Universitätsbibliothek* (Prague: Rohliček, 1851), p.274; *Zusätze,* ed. I. J. Hanuš (Prague, 1863), p.9.

17. F. J. Studnička, *Prager Tychoniana* (Prague: Verlag der. k. böhmischen Gesellschaft der Wissenschaften, 1901), p.43. Two other attributions to Brahe by Studnička were rejected by TB I, 315, 317.

amount of blank paper which is bound into the volume."[18] The point of view adopted by Studnička and Prusík was endorsed by Richard Kukula[19] (1857-1927), director of the Prague University Library from 1897 to 1918. Wilhelm Prandtl (1878-1956) agreed that Copernicus-plus-Rheticus has "numerous annotations by Brahe,"[20] a description concurred in by Flora Kleinschnitzová (1891-1946)[21] and Emma Urbánková.[22]

This array of students of Tycho and of the Prague University Library (Hanslik, Hanuš, Studnička, Prusík, Kukula, Prandtl, Kleinschnitzová, Urbánková) culminated in the 1971 Cimelia, which asserted, "A librarian in the Prague Jesuit College Library ... explicitly fixed ... the author of the commentary"[23] in Copernicus-plus-Rheticus as Tycho Brahe. While conceding that the Jesuit entry was "not entirely free of possible doubt,"[24] Cimelia came down on the affirmative side, maintaining that "Tycho Brahe is really the author of these notes. ... [W]e have the possibility of judging the handwriting. ... [E]ssentially it is like the writing in well-known and undoubted relics of Tycho's manuscripts."[25] But Cimelia did not document this asserted essential similarity by presenting handwriting samples of Tycho and the annotator side by side. Yet this very presentation had been made (inadvertently) in 1952 by *Astronomy in Czechoslovakia from its Early Beginning to Present Times,* pages 90 and 91.[26] Page 90 shows Copernicus-plus-Rheticus, folio 75r, with notes in all four margins.[27] Page 91 shows page 268 of the copy of an edition of Ptolemy (Basel, 1551) which was bought by Brahe in

18. B. Prusík, "Tychoniana der Prager k. k. Universitäts-Bibliothek," *Mittheilungen des österreichischen Vereines für Bibliothekswesen,* 5 (1901), 199/#VI. Cimelia counts "21 blank folios in front of the actual work and another 71 folios after it" (p.15/10↑-9↑).

19. P.24/10↑-7↑, in the work cited in n.11, above.

20. P.323/#14 (1932); p.11/#14 (1933), in the works cited in n.10, above.

21. F. Kleinschnitzová, "Ex bibliotheca Tychoniana Collegii Soc. Jesu Pragae ad S. Clementem," *Nordisk Tidskrift för Bok- och Biblioteksväsen,* 20 (1933), 86/#11.

22. P.57/#404 and plate 110, in the work cited in n.13, above.

23. Cimelia, p.12/24-27.

24. Cimelia, p.12/22-23.

25. Cimelia, p.12/11↑-6↑.

26. *Astronomie v Československu od dob nejstarších do dneška* (Prague: Státní osvětové nakladatelství národní podnik, 1952).

27. Better reproduction in Urbánková, plate 110.

Copenhagen for two dollars on 30 November 1560,[28] with his notes in the left margin. One need not be an expert paleographer to see at a glance that Tycho's handwriting is "essentially" different from the handwriting in Copernicus-plus-Rheticus. Yet this difference was not recognized by *Astronomy in Czechoslovakia,* which attributed to Tycho the annotations in Copernicus-plus-Rheticus.

Cimelia also claims that the "extensive commentary by Tycho Brahe makes it possible to judge in detail how Brahe's relation toward Copernicus's views was formed."[29] To test Cimelia's claim, let us look at the angle between the planes of the equator and the ecliptic, or "the obliquity of the ecliptic," as it is usually called. Copernicus says "in our time it is found not greater than $23°28\frac{1}{2}'$."[30] In Copernicus-plus-Rheticus, this value is repeated in the left margin by the annotator, who accepts a steady decrease in the obliquity of the ecliptic from $23°52'$ in the time of Aristarchus to $23°28'$ in his own time: "At this time [it is] $23°28'0''$." Brahe, however, states: "Now, after the passage of several years with the aid of many instruments and by the use of great care, I have found this [greatest declination of the sun or obliquity of the ecliptic] to be $23° 31\frac{1}{2}'$."[31] Evidently, the annotator is not Brahe. Nor is the annotator a member of Brahe's team who was familiar with Tycho's determination of the obliquity. Yet the annotator's copy of Copernicus-plus-Rheticus became part of Brahe's library, as we shall see later on.

In December of that year Professor Owen Gingerich of Harvard University called attention to a previously overlooked copy of the first edition of Copernicus's *Revolutions* in the Vatican Library, Fondo Ottoboniano Latino, #1902.[32] Although this is a printed book, it was grouped by the Vatican Library with the manuscripts on account of its supplementary handwritten sheets and numerous unsigned marginal notes. These were written by the same hand that was responsible for the Prague notes, while the contents of both sets of notes are pretty much

28. Zincograph in Studnička, p.38.

29. Cimelia, p.11/last 4 lines.

30. Copernicus-plus-Rheticus, fol.65v/9-10; p.122/30, in the work cited in n.1, above.

31. TB II, 18/7-9.

32. Paul Oskar Kristeller, *Iter Italicum,* II (London: Warburg; Leiden: Brill, 1967), 419/#1902; not #1901, as in Owen Gingerich, "The Astronomy and Cosmology of Copernicus," International Astronomical Union, *Highlights of Astronomy,* 3 (1974), 83. Hereafter cited as "Highlights."

in agreement. Hence, the annotator of the Vatican copy of the first edition of Copernicus's *Revolutions* is identical with the annotator of the Prague copy of Copernicus-plus-Rheticus.

Gingerich hastily accepted at face value Cimelia's recent identification of the Prague annotator as Tycho Brahe. Accordingly, Gingerich concluded that the Vatican annotations constituted "probably the most important Tycho manuscript in existence."[33]

Another point of similarity between the Vatican Copernicus and the Prague Copernicus-plus-Rheticus, in addition to the handwriting and the contents of the notes, is the binding of a considerable amount of paper behind the last printed page in each of these two volumes. These extra sheets remained blank in the Prague Copernicus-plus-Rheticus.[34] But the annotator of the Vatican Copernicus wrote on "30 manuscript pages at the end."[35] The first two of these thirty pages contain diagrams of the cosmos according to Copernicus, drawn with the sun at the center of the universe, and dated 27 January 1578.[36] Later on, however, in connection with the annotator's "Theory of the Three Outer Planets Adjusted to a Stationary Earth," he wrote: "This new system of hypotheses was discovered by me on 13 February 1578."[37] In those two and a half weeks the annotator shifted from a Copernican to a non-Copernican stance, from the earth as a planet in motion to a motionless earth.

Believing that the Vatican's is "Tycho's personal copy" of Copernicus's *Revolutions,* Gingerich maintained that "Tycho's notes show how he evolved his non-Copernican model."[38] Acknowledging that there is a chronological difficulty here, since Brahe "did not establish the Tychonic system until around 1583, five years after he drew these diagrams," Gingerich added: "I can only suppose that these five years were an important time of maturing"[39] from 1578 to 1583.

33. Highlights, p.81/18-19; Owen Gingerich, "Copernicus and Tycho," *Scientific American,* 229 (December 1973), 99/14↑-13↑. Hereafter cited as "C and T."

34. See the extract quoted above at n.18.

35. C and T, p.99/34-35; Highlights, p.80/4↑-3↑.

36. C and T, p.99/37-44; Highlights, pp.81-82. In Cimelia, folio 9v, below Copernicus's statement "At rest . . . in the middle of everything is the sun," the annotator wrote *focus Universi* (hearth of the universe), and he also applied this description to the sun in Copernicus's famous diagram of the cosmos.

37. Reproduced in C and T, p.90.

38. C and T, p.87/3-4.

39. C and T, p.101/11-16; Highlights, p.82/22-24.

In the Tychonic system, Brahe had the three outer planets (Saturn, Jupiter, Mars) orbit the sun, whereas they orbit the earth in the Vatican annotator's diagram, drawn on 17 February 1578.[40] Ten years later, in 1588, Brahe first published the Tychonic system,[41] which provoked a spate of correspondence, including a letter in which he looked back over his intellectual development from his acquiescing youth until he rejected the contemporary competing cosmologies. At that time he asked himself what "if the sun were established as the center of the five planets, and nevertheless ... revolved once a year around the earth, at rest in the center of the universe."[42] Thus, from the very beginning of his independent swing away from the prevailing conventional wisdom, Brahe centered all five planets on the sun. From this conception, as previously expounded by Copernicus, Brahe never deviated throughout the rest of his life. He first adopted this view some five years after the Vatican annotator had centered the three outer planets on the earth. Those five years were not "an important time of maturing" for Brahe, as Gingerich supposed. Instead, those five years measured the lag between the Vatican annotator's modification of Copernicus's cosmology in one direction and Brahe's somewhat later divergence in another direction.

Tycho Brahe is not the Vatican annotator any more than he is the Prague annotator. The Vatican handwriting is not his; its non-Copernican stance emerges before his; and its three outer planets are earth centered, not sun centered, like his.

Since Tycho Brahe was neither the Vatican annotator nor the Prague annotator, who was? Whoever he was, he knew the Copernican astronomy well, so well that he could try to revise it. His annotated copy of Copernicus-plus-Rheticus was acquired by Brahe. These clues point to Paul Wittich of Wrocław, whose name was mentioned to me by Professor Gingerich in a personal letter dated 31 July 1980.

In 1582-84 Wittich gave private instruction to Duncan Liddel (1561-

40. Four days after the annotator's shift to a non-Copernican stance on 13 February 1578; not "two days later," as in C and T, p.99/7↑, nor "three days later," as in Highlights, p.82/7. The drawing of 17 February 1578 is reproduced in Edward Rosen, "Render Not unto Tycho that which Is Not Brahe's," *Sky and Telescope*, 61 (1981), 476. Afterwards cited as "Render."

41. TB IV, 155-70; partial English translation by Marie Boas and A. Rupert Hall, "Tycho Brahe's System of the World," *Occasional Notes of the Royal Astronomical Society*, 3 (1959), 257-63.

42. TB VII, 129/40-130/2.

1613), who later became the "first in Germany to teach the theories of the heavenly motions according to the hypothesis of Ptolemy and Copernicus at the same time."[43] In the winter semester of 1579 Liddel, a Scot from Aberdeen, had enrolled in the University of Frankfurt on the Oder,[44] where his fellow countryman John Craig (?-1620) was then teaching. The "first principles" of Copernicanism were imparted by Craig to Liddel, who "learned more completely from Wittich ... about Copernicus's innovative hypotheses, which with no impropriety are regarded as marvelous."[45]

"Once," as Brahe wrote from his observatory to Háyek on 4 November 1580, "when I went to [the University of] Wittenberg, Wittich was a student there and he discussed with me the pursuit of these studies [looking to the simplification of astronomical calculations]. Stimulated by the remarks which he heard from me, he seemed, as he freely admits, to have taken only certain first steps here."[46] For, after Brahe and Háyek had become friends when they met at the coronation of Rudolph II at Regensburg on 1 November 1575, their efforts to communicate with each other were ineffective until the summer of 1580, when Háyek's letter recommending Wittich to Brahe was brought to Tycho in person by Wittich. "But when he had stayed with me a little over a quarter of a year and thought he had already acquired enough of what he wanted, he alleged that his uncle in Wrocław had died, whose legacy would bring him great profit if he were there in time. Accordingly, having readily obtained permission to leave, he promised to come back here within seven or eight weeks."[47] But Wittich never returned to Brahe.

During Wittich's brief stay in Brahe's observatory on the island of Hven in the Danish Sound, Tycho first noticed the comet of 1580 on 10 October. On 20 October he went to Hälsingborg on the mainland for

43. Duncan Liddel, *Ars medica* (Hamburg, 1607-08, 1617, 1628), letter of 1 May 1607 from Johannes Caselius to John Craig.

44. *Aeltere Universitäts-Matrikeln*, I, Universität Frankfurt a. O., ed. Ernst Friedlaender (Leipzig: Hirzel, 1887; rpt. Osnabrück: Zeller, 1965; Publicationen aus den k. Preussischen Staatsarchiven, xxxii), 277/#75. Wittich had enrolled in the summer semester of 1576 as a master of arts (p.253/1-3, where he is designated by a later hand as Rudolph II's Imperial Mathematician). Craig, a Scot from Edinburgh, was enrolled as an M. D. and member of the faculty in the summer semester of 1573 (p.228/#23).

45. Liddel, *Ars medica*, Caselius to Craig.

46. TB VII, 58/36-39. Brahe first met Wittich at Wittenberg toward the end of 1575.

47. TB VI, 89/24-29.

about a week. During his absence from Hven the comet was observed there by Wittich and an associate, who "made the following observations on Hven on the same days [as Tycho in Hälsingborg] until the 28th, when I returned home," Brahe reports.[48] The Hven cometary observations of 21, 22, and 26 October 1580 were recorded by Wittich.

These pages marked with the letters ABCDE, I recognize to have been written by the hand of Paul Wittich, of blessed memory, which is very well known to me, and I so testify with this handwritten note which I left at Prague with the magnificent and most noble lord Tycho Brahe on 23 October 1600.

Jacob Monaw
with my own hand[49]

Wittich's handwriting, a specimen of which is reproduced from his cometary observations,[50] differs from Brahe's[51] and matches the annotations in the Prague Copernicus-plus-Rheticus and those in the Vatican *Revolutions* of Copernicus. Wittich, not Tycho Brahe, is our Prague and Vatican annotator.

On 1 May 1582 Háyek wrote to Brahe from Prague: "During this past Lenten season I was with my daughter at Wrocław, and I stayed there more than a month. I discussed my [conclusions about the] parallaxes with Wittich, who gave me your observations of the recent comet which you made on 11, 12, 13, and 17 October[52] [1580]." Wittich died on 5 January 1586.[53] A dozen years later, on 24 March 1598, Brahe wrote to a

48. TB XIII, 315/15-16. On the following day, 29 October 1580, Wittich left Hven. As a parting gift, Brahe presented him with a copy of Peter Apian's *Astronomicum caesareum* (Ingolstadt, 1540). This is now in the Regenstein Library, University of Chicago. A photocopy of the title page showing the presentation was kindly supplied by Professor Martin J. Hardeman, Roosevelt University.

49. TB XIII, 316/n.1; English paraphrase by J. L. E. Dreyer, "On Tycho Brahe's Manual of Trigonometry," *Observatory*, 39 (1916), 129-30.

50. TB XIII, 317/upper left corner.

51. Wittich's *Occasus, aquila,* and *Informis* may be compared with Brahe's (TB XIII, 308, 319). Three examples of Brahe's handwriting taken from his comet notebook (not "treatise") are reproduced in Render, p.477.

52. TB VII, 72/7-11.

53. *Dictionary of Scientific Biography* (New York: Scribner's, 1970-80), XIV, 470/5↓, dates Wittich's death on 9 January 1587, following Rudolf Wolf, "Beiträge zur Geschichte der Astronomie. 3. Paul Wittich aus Breslau," *Vierteljahrsschrift der astronomischen Gesellschaft,* 17 (1882), 129/12↓. Wolf cited *Silesia togata,* an unpublished manuscript by Nicholas Henel of Hennefeld (1584-1656). But on fol.415v Henel wrote 6 over a

favorite pupil: "A few days ago my very dear friend Jacob Monaw of Wrocław wrote to me that at the end of last year [1597] you came to Wrocław.... In the same letter Monaw reported that he arranged to have you introduced to the sister of Wittich, of blessed memory, where you examined all his books. I therefore ask you to inform me about them, what they were and of what sort, especially the manuscripts, and whether she wants to sell them and at what price."[54] Nearly two years later Monaw (1546-1603) wrote to Brahe on 14 March 1600:

As far as Wittich's books are concerned, I want you to know that at the time when your letter was delivered here, Paul Wittich's sister, already advanced in age, celebrated her second marriage, of which the result was that the newly-wed died after sixty days. She left one son as her heir, for whom legitimate guardians have not yet been appointed. When this is taken care of, I shall see whether anything can be done with them.... I hear from those who examined Wittich's books while his sister was still alive that everything is mixed up and confused in various ways, like the Sybilline Books, and I am not surprised. She was a very strange woman, full of her own ideas.[55]

If Monaw succeeded in reaching an understanding with the guardians of the nephew of Paul Wittich, his annotated copy of Copernicus-plus-Rheticus may have been acquired by Monaw for Tycho Brahe. It would then have passed through George Brahe's heirs to the Prague Jesuits and thence to the library of Prague University, whence it issued in 1971 in facsimile as Cimelia, designated as the "Copernicus-Brahe *De revolutionibus*." Cimelia's author was Copernicus; its second owner was Tycho Brahe; and its original owner and annotator was Wittich.[56]

A copy of the first edition of Copernicus's *Revolutions* is preserved in the Saltykov-Shchedrin National Public Library in Leningrad.[57] Ac-

---

previous 5, faintly suggesting a 7 to Wolf's informant, who also misread the day 5 as 9. The library of the University of Wrocław kindly provided a microfilm of Henel's discussion of Wittich.

54. TB VIII, 34/4-16.

55. TB VIII, 266/1-11.

56. When Tycho wrote his *Reply to the Scot [John] Craig about Comets, A Defense,* in 1589, he praised Wittich's "understanding of Copernicus" (TB IV, 453/6-7).

57. N. I. Nevskaia, "A Unique Copy of Copernicus's Book," *Priroda,* 1978, 141; *Voprosy istorii estestvoznania i tekhniki,* 61-63 (1979), 83-84; *Isis,* 71 (1980), Critical Bibliography, 108/#1453.

cording to Gingerich, this copy was annotated by several astronomers, including Brahe, with some notes in his own hand. But after his return to the United States, Gingerich recognized from the microfilm of the Leningrad copy that "not a single one of the annotations recalls the handwriting of Tycho Brahe."[58] Like the Copernicus *Revolutions* (Ottoboniano Latino #1902) in the Vatican, and the Wittich copy in Prague, the Leningrad copy of the *Revolutions* was not annotated by Tycho Brahe.

58. O. Gingerich, "Annotations in Copernicus's Book," *Voprosy* ... (1980), 105/1.

22

Kepler and the Lutheran Attitude towards Copernicanism

As WE honor the memory of that great genius Johannes Kepler, who was born 400 years ago in 1571, we would do well to recall how he first became acquainted with the Copernican astronomy, and then strove to persuade his fellow-Lutherans in particular, and mankind in general, to accept Copernicus' new and revolutionary cosmology.

After preparatory study elsewhere, on 18 October 1589 Kepler was officially admitted to that leading Lutheran center of higher learning, Tübingen University.[1] There he had the good fortune to study astronomy with Michael Mästlin (1550–1631), who in 1590 gave him the grade of A on 20 January, A− on 23 April, and A on 22 July. However, it was not these grades, which were just like the grades received by Kepler in his other subjects, that especially endeared Mästlin to him. On the contrary, Kepler owed a lifelong debt to Mästlin for introducing him to the ideas of Nicholas Copernicus (1473–1543), the founder of modern astronomy. In the preface to his first major work, the *Mysterium Cosmographicum*,[2] Kepler wrote in 1596:

---

[1] Edmund Reitlinger, C. W. Neumann and C. Gruner, *Johannes Kepler* (Stuttgart, 1868), p. 210.

[2] *Mysterium Cosmographicum*, Tübingen, 1596; translated into German by Max Caspar, *Johannes Kepler, Das Weltgeheimnis* (Augsburg, 1923; Munich/Berlin, 1936).

"When I was studying with the highly renowned Professor Michael Mästlin at Tübingen six years ago ... Mästlin mentioned Copernicus in his lectures very frequently."[3] (Fig. 1)

The foregoing passage was translated from Latin into German by the late Max Caspar (7 May 1880–1 September 1956) in his biography of Kepler.[4] Although Caspar quoted Kepler's statement that "Mästlin mentioned Copernicus in his lectures very frequently", on the same page in his biography Caspar declared that Mästlin "reported ... on Copernicus' cosmological doctrines only ... in a more intimate circle".[5] Mästlin's lectures, however, were delivered to all the students in his regular class, and not to a "more intimate circle".[6]

Caspar also maintained that Mästlin, "the astronomy professor, sided completely with the system of Ptolemy's *Syntaxis* in his public lectures as well as in all the editions of his *Epitome*"[7] *of Astronomy* (Heidelberg, 1582; Tübingen, 1588, 1593, 1597, 1598, 1610, 1624). However, the *Epitome* was an elementary textbook, in which Mästlin refrained from expressing his true convictions. As he explained to his ruler, Duke Frederick of Württemberg (1577–1608), "To be sure, in all schools the familiar ancient hypotheses are retained for the young and, being easier to understand, are taught as correct; but all specialists as a body agree with Copernicus' demonstrations."[8] Unlike his introductory pedagogical textbook, Mästlin's classroom lectures at Tübingen University revealed his professional attitude toward Copernicanism. As Kepler said, "partly from Mästlin's utterances, partly by my own efforts, I gradually made a collection of the superiorities of Copernicus over Ptolemy from the mathematical point of view".[9] Since Kepler compiled his catalogue of Copernican superiorities over Ptolemy "partly from Mästlin's utterances", it is obviously not true that Mästlin "sided completely with the system of Ptolemy's *Syntaxis*". In short, we have Kepler's incontrovertible testimony that at Tübingen University in 1590 Mästlin not only mentioned Copernicus in his classroom lectures to his students, but also pointed out to them superiorities of Copernicus over Ptolemy.

Kepler remarked that he "could easily have dispensed with the labor" of compiling the catalogue of those superiorities, since Copernicus' disciple, "George

[3] Johannes Kepler, *Gesammelte Werke*, cited hereafter as "*GW*" (Munich: Beck, 1937–), I.9: 11–14.

[4] Max Caspar, *Johannes Kepler* (Stuttgart: Kohlhammer, 1948; 2nd ed., 1950; 3rd ed., 1958).

[5] Caspar, *Kepler*, p. 48.

[6] Caspar's mistaken restriction of Mästlin's audience was applied to Tübingen as a whole by Gerhard Kropatscheck, *Johannes Kepler* (Stuttgart, 1947; Begegnungen, Heft 5), p. 10.

[7] Caspar, *Kepler*, p. 48.

[8] *GW*, XIII, 68: 45–48.

[9] *GW*, I, 9: 20–21; Caspar, *Kepler*, p. 49. In *GW*, I, which was edited by Caspar a decade before he wrote his biography of Kepler, he came closer to the historical truth when he said that Mästlin "set forth the superiority of the new Copernican doctrine" and "dared to express his agreement with the new doctrine". However, in both these passages (*GW*, I, 403, 413) Caspar mistakenly restricted Mästlin's utterances to "his more trusted students" and "most trusted circles".

MICHAELIS MÆSTLINI. GOEP
PINGENSIS, NATI ANNO 1550·30·
SEPTEMB·MATHEM·IN INCLYTA·
TVBING: ACADEMIA AB ANNO 1584·
PROFESSORIS EFFIGIES ANNO 1610

FIG. 1. Michael Mästlin (1550–1631). (Landesbildstelle Württemberg, Stuttgart.)

FIG.   2. Christoph Besold (1577–1638). (Landesbildstelle Württemberg, Stuttgart.)

FIG.   3. Veit Müller (1561–1626). (Landesbildstelle Württemberg, Stuttgart.)

Joachim Rheticus[10] . . . [had] pursued the individual [comparisons] briefly and clearly in his *First Report*".[11] Hence, this earliest printed account of the new astronomy, which had been published in 1540, was available to Kepler at Tübingen. Although Kepler's comment about Rheticus had been translated by Caspar in 1923, in his 1948 biography Caspar said that Kepler "at the time of his earliest efforts did not even . . . know . . . Rheticus' *First Report*".[12]

Moreover, in his 1948 biography Caspar declared: "In any case at that time he [Kepler] did not get to read Copernicus' work itself."[13] Yet in 1923 Caspar had translated Kepler's reference in 1595 to a certain intention which he had "always had in mind from the time when [he] began to examine the [six] books of Copernicus' *Revolutions*" of 1543.[14] Moreover, on 3 October 1595 in a letter to Mästlin, Kepler enumerated three reasons why he had "always sided with Copernicus".[15] Unfortunately, in 1595 Kepler did not specify exactly when he "began to examine . . . Copernicus' *Revolutions*". But how could he have compiled his list of Copernicus' superiorities over Ptolemy without consulting the *Revolutions*?

True, in that connection he mentioned only Rheticus and omitted Copernicus. The latter, however, for the most part studiously avoided claiming any superiority over Ptolemy. Rheticus, on the other hand, inflamed with the fiery enthusiasm of the fervent disciple, did not feel bound by the restraints which Copernicus imposed on himself. Hence Kepler's failure to couple Copernicus with Rheticus as potential sources for his catalogue of the superiorities of the new astronomy over Ptolemy does not indicate that Kepler lacked access to Copernicus' *Revolutions* while he was a student at Tübingen.

In point of fact Kepler possessed his own personal copy of the *Revolutions*,[16] concerning which Caspar said that Kepler "had [it] in his possession later".[17] Although Caspar's expression "later" is deliberately vague, since we do not know precisely when Kepler acquired his copy of the *Revolutions*, Caspar unquestionably meant "later" than Kepler's student years at Tübingen, which extended from 1589 to

---

[10] Karl Heinz Burmeister, *Georg Joachim Rhetikus* (Wiesbaden: Pressler, 1967–8).

[11] *GW*, I. 9: 21–23; *Weltgeheimnis*, p. 19. Rheticus' *Narratio prima* was translated into German by Karl Zeller, *Des Georg Joachim Rheticus Erster Bericht* (Munich/Berlin, 1943), and into English by Edward Rosen, *Three Copernican Treatises*, 3rd ed. (New York: Octagon, 1971).

[12] Caspar, *Kepler*, p. 49.

[13] *Loc. cit.*

[14] *GW*, I. 14: 26–27; *Weltgeheimnis*, p. 29; the nature of Kepler's intention is clarified in the text at note 52, below.

[15] *GW*, XIII. 34: 46. Kepler's past tense (*adhaeserim*) was missed in the German translation by Max Caspar and Walther von Dyck, *Johannes Kepler in seinen Briefen* (Munich/Berlin, 1930), I. 18.

[16] A facsimile of Kepler's copy of the *Revolutions* was reprinted by Johnson (New York/London, 1965) and Edition Leipzig.

[17] Caspar, *Kepler*, p. 49. In his letter of 3 October 1595 to Mästlin, Kepler referred to "my copy" of Copernicus' *Revolutions*, which he had evidently been studying for some time (*GW*, XIII, 45: 455).

1594. In his copy of the *Revolutions*, on the blank leaf preceding the title-page, with his own hand Kepler wrote a brief dialogue in poetic form, which he dated "22 December: 98". [17a] This entry, however, tells us only that the copy was in Kepler's possession on 22 December 1598, without informing us how long it had been in Kepler's possession before that date. Whenever Kepler acquired his own copy of the *Revolutions*, in 1590 at Tübingen Mästlin based his judgement concerning the superiorities of Copernicus over Ptolemy on the copy of the *Revolutions* which he had acquired 20 years before. [18]

Finally, in his biography of Kepler, Caspar contended that Mästlin expounded Copernicus "only with prudent reserve". [19] Yet Caspar had translated Kepler's declaration that he "joined this party" of the Copernicans "not . . . without the extremely weighty authority of my teacher Mästlin, a most renowned astronomer". [20]

If we now assemble the separate pieces of our composite picture of Kepler and Mästlin at Tübingen, we notice a startling contrast between Caspar's view and Kepler's. According to Caspar's biography, as a student at Tübingen Kepler had access neither to Copernicus nor to Rheticus, while Mästlin spoke about Copernicus only in a more intimate circle, with prudent reserve, and sided completely with Ptolemy. On the other hand, the *Mysterium Cosmographicum* shows us Kepler having access to Copernicus' *Revolutions* as well as to Rheticus' *First Report*, while embracing Copernicanism "not without the extremely weighty authority of Mästlin", who often mentioned Copernicus in his classroom lectures and pointed out his superiority over Ptolemy. [21]

In the preface of his *Mysterium Cosmographicum* Kepler said that when he was Mästlin's student, "I was so delighted with Copernicus . . . that I often defended his opinions in the students' debates about physics. I also wrote a painstaking disputation about the first motion [that is, the apparent daily rotation of the entire heavens], contending that it happens on account of the rotation of the earth." [22] Many years later, addressing Christopher Besold (1577–1638) (Fig.   2) who had been a younger fellow-student at Tübingen, in his posthumous *Somnium* Kepler said:

> I have a very old document which you, most illustrious Christopher Besold, wrote with your own hand when, in the year 1593, on the basis of my essays you formulated about twenty theses concerning the celestial phenomena on the moon and showed them to Veit Müller,

[17a] Kepler's poem was translated into German and analyzed by Friedrich Seck, "Ein Gedicht von Johannes Kepler," in *Staatsanzeiger für Baden-Württemberg, Beiträge zur Landeskunde*, No. 3–4, pp. 8–12 (Oct. 1966).

[18] Mästlin's copy of Copernicus' *Revolutions*, which is still preserved in the Schaffhausen Municipal Library, was used extensively by him in his 1573 essay on the new star of 1572 (*Tychonis Brahe dani opera omnia*, III, 58–62, Copenhagen, 1913–29).

[19] Caspar, *Kepler*, p. 48.

[20] *GW*, I, 16: 39–40; *Weltgeheimnis*, p. 31.

[21] As was stated entirely correctly by Reitlinger, p. 128.

[22] *GW*, I, 9: 13–17; *Weltgeheimnis*, p. 19.

who then regularly presided over the philosophical disputations, with the thought that you would engage in a debate over them if he approved.[23]

But Veit Müller (1561–1626), professor of philosophy at Tübingen University (Fig. 3), did not approve, as Kepler plainly implies here. Müller's adamant adherence to the traditional astronomy was mentioned explicitly by Kepler in his aforementioned letter to Mästlin of 3 October 1595: "Professor Müller showed this [attitude] even in the disputation that I wrote about the moon, which Besold was otherwise prepared to defend."[24] The presence of "otherwise" (*alioqui*) was overlooked by Caspar, who mistakenly asserted that "Professor Veit Müller presided" over Kepler's lunar dissertation of 1593.[25]

Veit Müller refused to preside over the Kepler–Besold lunar disputation because he was opposed to the Copernican cosmology. The distinctive feature of that astronomical system was its insistence that the Earth moves. On the other hand, a common objection to Copernicanism was the argument that if the Earth really does move, it should provide its inhabitants with some perceptual evidence of its motion. The refutation of this common anti-Copernican argument was the principal purpose of the Kepler–Besold lunar theses of 1593. Although the theses themselves have not survived, their basic contents are known from Kepler's later references to them. In describing how the appearances in the heavens would look to imaginary inhabitants of the Moon, the 1593 theses reasoned that such lunar creatures would suppose the Moon to be stationary, since they shared in its motions. Earth-dwellers, by contrast, see perfectly plainly that the Moon actually does move. These same earth-dwellers lack empirical evidence of their own planet's motion only because they participate in that motion. Some such analysis formed the substance of the 1593 lunar theses, which Müller would not allow Besold to debate. This rejection was the individual decision of Müller, to whom had been assigned the task of presiding over Tübingen's philosophical disputations. The Kepler–Besold matter did not go beyond him to the other members of the Tübingen faculty.

A little more than 2 years before, on 11 August 1591, Kepler had been awarded a master's degree.[26] He thereupon entered the theological course, which was scheduled to last 3 years. Half-way through his third and last year as a student of theology, however, an event occurred which completely altered the plan of his life. George Stadius, the teacher of mathematics at the Lutheran school in Graz, died and the local authorities asked Tübingen to send a replacement. Kepler was selected,

---

[23] Kepler, *Somnium*, facsimile of the 1634 edition, eds. Martha List and Walther Gerlach (Osnabrück: Zeller, 1969), p. 29; at p. 32 in the English translation by Edward Rosen, *Kepler's Somnium* (University of Wisconsin Press, 1967), awarded the Charles Pfizer Prize in 1968 by the History of Science Society of the United States.

[24] *GW*, XIII. 39: 242–4.

[25] *GW*, XIII. 379, no. 23, line 243.

[26] *Die Matrikeln der Universität Tübingen* (Stuttgart/Tübingen, 1906–54), I. 655.

and in accepting the post he subordinated his hope of becoming a Lutheran clergyman, and instead entered upon the career which was destined to bring him immortal renown as one of mankind's greatest scientists.

While teaching at Graz, Kepler composed his *Mysterium Cosmographicum*, which he submitted to George Gruppenbach,[27] a publisher in Tübingen, when he returned there on a visit. Gruppenbach pronounced himself prepared to proceed, provided that the academic authorities at Tübingen University would approve.[28] Accordingly on 1 May 1596 Kepler addressed a letter to the rector, deans, and professors of the university, "humbly submitting ... his manuscript ... to their scrutiny", and asking for their approval "if nothing in my subject is injurious to the academic community".[29]

The later developments will be properly understood only if it is constantly borne in mind that as a young impecunious teacher, convinced that he had made a momentous cosmological discovery, Kepler submitted the manuscript of his *Mysterium Cosmographicum* to the scrutiny of Tübingen University because the publisher would go ahead only with the "knowledge and approval" of the university authorities.[30] They were in a position to exercise censorship over Kepler's *Mysterium Cosmographicum* and to require him to make changes in it only because of Gruppenbach's insistence that Kepler should obtain the Tübingen faculty's prior approval of his book.

The Tübingen faculty naturally voted to have Kepler's cosmographical manuscript examined by Mästlin, who was the professor of mathematics and astronomy. A copy of his highly favorable report, which was unanimously endorsed by the University Senate, was sent to Kepler.[31] He was asked, however, to alter his manuscript in two ways before Gruppenbach's press would be permitted to start operating. First, all obscure passages were to be clarified to the best of his ability. Secondly, by way of a preface he was to explain "the hypotheses of Copernicus".[32]

It cannot be too strongly emphasized that the Lutheran authorities at Tübingen University, far from compelling Kepler to suppress or conceal his adherence to Copernicanism, required him to compose additional material expounding the new

---

[27] Josef Benzing, *Die Buchdrucker des 16. und 17. Jahrhunderts im deutschen Sprachgebiet* (Wiesbaden: Harrassowitz, 1963; *Beiträge zum Buch- und Bibliothekswesen*, Bd. 12), pp. 436–7. Gruppenbach's inventory, dated 15 May 1597, does not list Kepler's *Mysterium Cosmographicum*, of which the author had agreed to take 200 copies (*GW*, XIII. 111: 153–7; 118: 216–19; Theodor Schott, "Zur Geschichte des Buchhandels in Tübingen: Lagerverzeichnis des Tübinger Buchhändlers Georg Gruppenbach vom Jahre 1597", *Archiv für Geschichte des deutschen Buchhandels*, 1879, 2. 244–51).

[28] *GW*, XIII, no. 40: 6–10; Caspar–Dyck, I. 31.

[29] *GW*, XIII, no. 40: 13–15; Caspar–Dyck, I. 31–32.

[30] *GW*, XIII, no. 40: 9; Caspar–Dyck, I. 31.

[31] *GW*, XIII, no. 44: 2–6.

[32] *GW*, XIII, no. 44: 9–13.

revolutionary astronomy. This requirement was based on a recommendation in the report by Mästlin, who said:

> This little book was written by Kepler, a very learned and most brilliant man. By prolonged concentration he has made himself completely familiar with the entire subject. Then he proceeds to assimilate his own mind to the minds of others, as though this treatise of his were to get into the hands only of those persons who have a thorough knowledge of all of Copernicus' demonstrations, even the most abstruse. . . .[33] Accordingly he either omits or states too concisely Copernicus' theory about the order of the spheres and their dimensions . . . as topics which are obvious and known to everybody. But the situation is entirely different. For of the readers who are going to peruse this book in order to enjoy its originality, not a few are devoted to other studies or arduous occupations. Even if [copies of] both Copernicus and Euclid are in their hands, they cannot take the time to learn the method or acquire the ability of mastering in those authors what is needed by way of preparation to understand Kepler. Even those who know Copernicus' writings well and are familiar with Euclid's theorems do not instantly have at their fingertips what is referred to here [in Kepler's book]. Consequently it will reach many persons without benefiting them, even though they may love mathematics deeply (I am not speaking of the fault-finders).
>
> I therefore wonder whether Kepler should be advised to expound his praiseworthy and notable discovery in a somewhat clearer and more popular style. This could perhaps be done if he wrote a preface (for the manuscript in its present form lacks a beginning) explaining Copernicus' hypotheses systematically and more fully. Kepler might review the order and sizes of the [celestial] spheres, and cite the passages in which their dimensions are treated by Copernicus. At the same time Kepler might provide a diagram and numerical tabulations, because the subject is absolutely incomprehensible without a diagram. This could be made simply, however, and in an ordinary manner, as it is in Copernicus, whose diagram is quite elementary yet satisfactory for his purpose.[34]

Mästlin concluded that Kepler's manuscript, after the minor modifications indicated above, was "eminently worthy of being published and thereby becoming available to educated men".[35] In this recommendation the entire Senate of Tübingen University concurred. Without any dissent they gave their approval to the publication of a book that was wholeheartedly committed to Copernicanism. What is more, they insisted that Kepler should make his book even more Copernican than it had been in the form submitted to them for their scrutiny. In the face of the Tübingen University Senate's unanimous endorsement of Copernicanism, as expounded by Kepler and recommended by Mästlin, the silence of Veit Müller in 1596 stands in significant contrast to his disapproval of the Kepler–Besold lunar theses of 1593 over which he had had sole authority.[36] Moreover, it was the prorector of the institution, the theologian Matthias Hafenreffer (1561–1619), who transmitted to Kepler Mästlin's enthusiastic eulogy of the militantly pro-Copernican *Mysterium*

---

[33] German translation in *Weltgeheimnis*, p. xxiv.

[34] *GW*, XIII, 85: 36–63; German translation in Caspar-Dyck, I. 38–39.

[35] *GW*, XIII, 84: 5–6; *Weltgeheimnis*, p. xxiii; Caspar-Dyck, I. 38.

[36] Müller's disapproval of the theses did not develop into any hostility toward Kepler as a person. For in a letter to Mästlin, Kepler said that Müller "no less than the others obtained for me this position" as teacher of mathematics at Graz (*GW*, XIII, 20: 33–34).

Cosmographicum, together with the University Senate's unanimous approval.[37] Hence we may reject as unhistorical Caspar's conclusion that while Kepler was a student, his allegiance to Copernicanism "cast a shadow on his otherwise excellent qualities and achievements in the eyes of his professors of theology".[38]

In those days a printed book customarily contained congratulatory poems or prose eulogies of the author and his work by his admirers. With regard to the possibility of obtaining such desirable adornments of his *Mysterium Cosmographicum*, in his aforementioned letter of 3 October 1595 to Mästlin, Kepler expressed his fear that, apart from three students of noble birth who were free, "the others under oath to the university would be unwilling to contribute anything opposed to the traditional opinion".[39] And in fact, apart from a short poem of eight lines, in the *Mysterium Cosmographicum* Kepler was lauded only by Mästlin.

After the publication of the *Mysterium Cosmographicum*, on 9 April 1597 Kepler told Mästlin how relieved he was that "the defenders of Holy Writ have raised no objection to my book, as I feared" that they would.[40] Kepler added: "I know of nobody in that most famous and illustrious university of yours who would assail my theses with malice aforethought."[41] Whereas in far-off Graz Kepler had heard of no scriptural opponent and no intentional antagonist, Mästlin on the spot was in a better position to detect the latent hostility. On 30 October 1597 he wrote to his former pupil that the *Mysterium Cosmographicum*

> somewhat offends our theologians too. They make no overt move, however, because they are deterred by the prestige of our duke, to whom the key diagram is dedicated. [In the *Mysterium Cosmographicum*, Plate III, exhibiting Kepler's conception of the universe as a nest of the six Copernican planets separated by the five regular solids, was dedicated to Duke Frederick, who had authorized the construction of a model.][42] Time and again Dr. Hafenreffer has assailed me (jokingly, to be sure, although serious tones too seem to be intermingled with the jests). He wants to debate with me, while defending his Bible etc. By the same token not so long ago in a public evening sermon he expounded Genesis, Chapter I: "God did not hang the sun up in the middle of the universe, like a lantern in the middle of a room" etc. However, I usually reply humorously to those jokes as long as they remain jokes. If the matter were to be treated seriously, I too would respond differently. Dr. Hafenreffer acknowledges your discovery to be wonderfully imaginative and learned, but he regards it as completely and unqualifiedly in conflict with Holy Writ and truth itself. Yet with these men (who are otherwise fine and very scholarly, but) who have no adequate grasp of the fundamentals of these subjects, in like manner it is better to act jokingly while they accept jokes.[43]

[37] *GW*, XIII, no. 44: 2–7.

[38] *GW*, I. 403.

[39] *GW*, XIII, 39: 241–2.

[40] *GW*, XIII, 113: 3–4; Caspar–Dyck, I. 44.

[41] *GW*, XIII. 116: 146–148; Caspar–Dyck, I. 50.

[42] *GW*, XIII, 69: 88–89. The duke was undoubtedly influenced by the unqualified statement in Kepler's letter to him of 29 February 1596 that "all the famous astronomers of our time follow Copernicus instead of Ptolemy and Alfonso", the patron of the Alfonsine Tables (*GW*, XIII, 66: 16–17).

[43] *GW*, XIII, 151: 19–32.

Mästlin's attitude shows unmistakably how erroneous was Christian Frisch's impression that Mästlin, "himself a theologian, seems to have been influenced by the authority of the theologians to say one thing while believing another".[44]

In replying to Mästlin on 6 January 1598 Kepler recalled "having changed the beginning" of the *Mysterium Cosmographicum* by inserting a preface

> on the advice of Hafenreffer, the proponent of the printing and eloquent encomiast of the discovery. That is why I wrote to him [this letter from Kepler to Hafenreffer has not survived], concealing what I know from your letter [of 30 October 1597]. And I really cannot believe that he is opposed to this doctrine [Copernicanism]. He pretends [to be so] in order to appease his colleagues, whom perhaps he displeased by furthering my book. And this he must be granted since for him peace with his colleagues is more important than peace with me.[45]

What Kepler told Mästlin more fully in the foregoing letter, he repeated more concisely in a letter to Hafenreffer (which has not survived). It reached the theologian (Fig. 4, p. 238) in February, and after a delay due solely to lack of leisure and couriers he replied on 12 April 1598 with a reference to Kepler's uncertainty

> whether to pass over the matter in silence or to try to make clear to everybody that those hypotheses of yours agree in all respects with Holy Writ. As I understood your letter [which has been lost, as was just mentioned], in that part of it you do not conceal your request for my brotherly advice. What I think about this subject, most illustrious sir and dearest brother, I shall reveal to you frankly and openly. Between those hypotheses and Holy Writ there must of course be a clear distinction. This I have always felt without any question, even from the time when I first saw the hypotheses, as you could observe when you were here with us [at Tübingen in May 1596] and can still remember perfectly well. For this is the reason why I recommended, not only in my own name but also in the name of my colleagues, the omission from your treatise of the chapter (I think it was No. 5)[46] which dealt with this harmonization [of Copernicanism with the Bible], lest those very [theological] disputes arise therefrom.[47]

From the period under discussion only two letters from Hafenreffer to Kepler survive, dated 6 June and 14 June 1596. In both of them Hafenreffer suggested changes in the manuscript of the *Mysterium Cosmographicum*, but without mentioning Chapter 5. However, toward the end of the second letter Hafenreffer remarked that he was writing briefly and would write "more fully at another time when he had leisure".[48] Perhaps it was in this later letter that Hafenreffer recommended the omission of Chapter 5. "This was indeed omitted", he continued in his letter of 12 April 1598 to Kepler, "but at the very beginning there is some mention, although quite

---

[44] *Vistas in Astronomy*, 18 (1975), Note A, p. 281.

[45] *GW*, XIII. 165: 127–33.

[46] Caspar (*GW*, I. 408) mistakenly said that what "fell as a sacrifice to the censor's blue pencil" was a portion of Chapter 1, whereas Hafenreffer's recollection involved the whole of Chapter 5. Without calling attention to Caspar's error in *GW*, I, in *GW*, VIII, 450, Franz Hammer in 1963 correctly stated that "a chapter" was deleted while Chapter 1 "was left intact".

[47] *GW*, XIII, 203: 24–35.

[48] *GW*, XIII, no. 48: 17.

brief, of this subject."[49] In Chapter 1 of the *Mysterium Cosmographicum* Kepler had been permitted to retain the opening paragraph, which reads as follows:

> It is an act of piety at the very beginning [*statim ab initio;* these three words were repeated by Hafenreffer] of this discourse about Nature to see whether it says anything contrary to Holy Writ. Nevertheless, I believe that it is premature to raise this question here before I am assailed.[50] In general I promise to say nothing harmful to Holy Writ, and if Copernicus is convicted of anything with me, I shall consider him finished.[51] And this was always my intention from the time when I began to examine the [six] books of Copernicus' *Revolutions.*[52]

This opening proclamation by Kepler that his *Mysterium Cosmographicum* would not conflict with the Bible had been allowed by Hafenreffer in 1596 to remain in the manuscript. After this draft had passed into the printer's hands, effective control over it by Hafenreffer and his fellow-theologians ceased. As he pointed out in continuing his letter of 12 April 1598 to Kepler,

> We cannot be blamed for the other material that was added later, especially the foreword[53] by M[ästlin], dated 1 October 1596, to Rheticus' *First Report,* which was appended to Kepler's *Mysterium Cosmographicum*], since none of these later additions were seen by us before they were sent to the printer. In the present situation, however,[54] I advise and caution you in a brotherly way not to try even to propound the aforementioned harmonization in public and to fight for it. For if you do, many good men might be incensed, not without cause, and the whole business might be blocked or bespattered with the noxious stain of dissension. For I have no doubt that if this opinion were openly advocated and defended, it would have its opponents, perhaps not lightly armed. Therefore, if there is any place for my brotherly advice (as I firmly hope), in demonstrating such hypotheses hereafter you will play the part of the pure mathematician, without worrying at all whether the hypotheses do or do not conform to the things in creation. For the mathematician accomplishes his purpose, I believe, if he produces hypotheses to which the phenomena correspond as closely as possible. And you yourself, I suppose, will yield to the man who could discover better hypotheses. Yet it does not follow that the truth of things instantly agrees with the hypotheses conceived by each and every specialist. I do not wish to adduce the irrefutable passages I could cite from Holy Writ. For in my judgment what is needed is not disputes but brotherly advice. If you heed it (as I confidently trust) and play the part of the abstract mathematician, I have no doubt that your thinking will be highly pleasing to very many persons (as it certainly is to me too). But if (may Almighty God forbid this) you should wish to harmonize those hypotheses with Holy Writ openly and fight for this harmonization, I am certainly afraid that this matter may result in dissension and strain. In that case I would wish that I had never seen your thoughts, although in themselves and considered mathematically they are splendid and lofty. For in God's [Lutheran] church there has long been more strife than is good for the weak.[55] However, I don't know where I am being carried along by my pen, or rather by my brotherly affection

---

[49] *GW*, XIII, 203: 35–37.

[50] German translation up to this point by Franz Hammer, *GW*, VIII, 450–1.

[51] If Caspar's mistranslation (*Weltgeheimnis*, p. 29) were right, Kepler would always have intended to disregard any charge that Copernicus was incompatible with the Bible. But Kepler means that if Copernicus were found guilty (not merely accused) of harming the Bible, he would cease to exist for Kepler.

[52] *GW*, I, 14: 21–27.

[53] *GW*, I, 82–85.

[54] Here begins a translation into German by Caspar, *Kepler*, pp. 74–75.

[55] Caspar's translation ends here.

for you. Were it not as very strong and very sincere as it is, I would not have permitted my pen such unfettered freedom. But this very affection makes two demands upon you: For us, play the role of the strict mathematician, and constantly foster that peace in the church which I know you have favored up to now.[56]

About 3 weeks after this heartfelt appeal from Hafenreffer, Kepler received a letter from Mästlin which he answered on 1/11 June 1598, in part as follows:

> You know what my purpose was in writing to Hafenreffer [the letter which the theologian received in February 1598, but which has been lost]. He replied, the gist of his message being as follows: I am to refrain from mentioning Holy Writ in public. [To do so] is bound to annoy many good men, among whom it is no secret that he numbers himself. Most of all he seems to be afraid that, buttressed by Duke Frederick's authority, I would wish through pressure to publish a defense [of Copernicanism] against the objections [based] on Holy Writ. For he says that I am going to find those who will want to stop the proceeding. In the meantime he urges me to press on vigorously with these [Copernican] hypotheses as far as they benefit astronomy. I now really believe much more what I had previously thought: the man is not opposed to Copernicus, but among the other theologians he must stand up for (what they think is) the Bible's authority. Therefore he does not tell me his true opinion.[57] What are we to do? The whole of astronomy is not important enough to offend any of Christ's little ones. Since most people, even those who are educated, do not use their intellect to rise to this lofty height of astronomy (as it seems to ordinary persons), let us imitate the Pythagoreans even in their customs. If anybody approaches us privately, let us tell him our opinion frankly. Publicly let us be silent. Why go and ruin astronomy by means of astronomy? The whole world is full of men who are ready to throw all of astronomy, if it sides completely with Copernicus, off the earth, and to forbid the specialists an income. But specialists cannot live off themselves or on air. Therefore, let us act in astronomical affairs in such a way that we hold on to the supporters of astronomy and do not starve. I say this because in writing me your latest letter but one [dated 30 October 1597], you remarked that you counter with jokes as long as Hafenreffer resorts to jokes, but if he should start a serious dispute with you, you are ready to treat him as he deserves. For I would not want you to make enemies or suffer setbacks for my sake or even for the sake of the truth itself (which affects nobody's salvation or income). Having observed that this man treats serious matters jokingly, you can drive out one nail with another and even when he treats serious matters jokingly, you can object seriously without engaging in combat in the name of another person. I say this bluntly, not copying you. You usually do not doubt that I shall do what you could urge me to do by virtue of your position. But if things look differently to you in this matter, I shall always be free to follow your advice. For at present I am not replying to Hafenreffer except that when the opportunity arises you may greet him with the utmost courtesy in my name.[58]

In his answer to Kepler on 4 July 1598 Mästlin said:

> With Hafenreffer I am initiating nothing which could be harmful. He has made many threats, albeit jestingly. However, the matter has remained undecided up to now. I answered jokingly because things have never become serious. That's the way they still stand.[59]

---

[56] *GW*, XIII, no. 93: 37–68.

[57] Here Kepler repeats his hunch about Hafenreffer, which he had previously expressed to Mästlin 5 months before, in the letter cited at note 45, above.

[58] *GW*, XIII, no. 99: 491–524; Caspar–Dyck, I. 78–80.

[59] *GW*, XIII, 236: 135–7.

Hafenreffer's appeal to Kepler clearly shows that the principal ingredient in the theologian's attitude toward the new astronomy was the fear that it might plant still another seed of discord among the Lutherans, who so sorely needed sincere unity of thought and action under the existing conditions of sharp rivalry with the other Christian denominations. That potential discord could be avoided, Hafenreffer reasoned, if Copernicanism were presented, not as a true picture of reality, but as a mere calculating device. Such a mathematical fiction could never conflict with the Bible, interpreted literally. If the authority of the Scriptures were not threatened by Copernicanism, its basic ideas would be unobjectionable, and even welcomed. As Kepler wrote to the Bavarian Chancellor, Hans Georg Herwart von Hohenburg (1553–1622), on 26 March 1598, "there is not an astronomer who puts these new hypotheses [of Copernicus] one bit behind those of antiquity; the struggle against Copernicus is waged exclusively and entirely by natural philosophers, metaphysicians, and theologians."[60]

Hafenreffer's attitude toward Copernican concepts as such had been indicated while he was supervising the revision of the manuscript of Kepler's *Cosmographic Mystery*. "With regard to the hypotheses of Copernicus" Kepler had proposed "to insert . . . the words of Rheticus."[61] But Hafenreffer objected to "Rheticus' [*First*] *Report* as too long-winded" and suffering from other defects.[62] Among those defects Hafenreffer did not include conflict with the Bible. The Tübingen theologian was not aware that Rheticus, an alumnus of the Wittenberg theological school, had written a "brief work most skillfully absolving the motion of the earth from disagreement with Holy Writ".[63] This was the judgment of the moderate Roman Catholic bishop Tiedemann Giese (1480–1550) concerning Rheticus' "brief work", which not surprisingly has perished.

The same fate did not befall the original Chapter 5 of the *Mysterium Cosmographicum*. This chapter was omitted from the printed book, it will be recalled, on the recommendation of Hafenreffer and his colleagues lest it give rise to theological disputes. Since the *Mysterium Cosmographicum* could not be published without the approval of the Tübingen faculty, Kepler perforce complied with their recommendation. But he did not discard the deleted chapter.

A clue to its contents may be found, I believe, in Kepler's aforementioned letter to Mästlin of 3 October 1595. At that time the *Mysterium Cosmographicum* had not yet been finished, and Kepler summarized its projected contents for his former teacher as follows: "At the outset in several theses I discuss Holy Writ, and I show

[60] *GW*, XIII, 193: 194–7. In Caspar–Dyck, I. 71, Kepler's *physicis* was mistranslated.

[61] *GW*, XIII, no. 47: 11–13.

[62] *GW*, XIII, no. 48: 5.

[63] Leopold Prowe, *Nicolaus Coppernicus* (Osnabrück: Zeller, 1967; reprint of the Berlin, 1883–4 edition), II, 420: 26–27.

how[64] its authority abides, and yet at the same time it cannot provide the basis for refuting Copernicus if otherwise he speaks reasonably."[65]

It will be noticed that in summarizing the *Mysterium Cosmographicum* for Mästlin on 3 October 1595 Kepler located his discussion of Holy Writ in several theses "at the outset" (*Initio*) of his manuscript. However, at the outset of the book as published in 1596, instead of a discussion of "Holy Writ in several theses" Kepler placed his "promise to say nothing harmful to Holy Writ". It would appear that after 3 October 1595 Kepler moved his "discussion of Holy Writ in several theses" to Chapter 5, which was then deleted to please Hafenreffer and his colleagues.

After being deleted, Chapter 5 was not discarded, as was indicated above. In later years, when Kepler was no longer an obscure teacher subject to censorship exercised by the Tübingen faculty, but the Imperial Mathematician of the Holy Roman Emperor Rudolph II, he incorporated what had been Chapter 5 of the *Mysterium Cosmographicum* into the Introduction to his *New Astronomy* (Heidelberg, 1609).[66] In that eloquent proclamation of the truth of the Copernican astronomy it is easy to recognize the "discussion of Holy Writ in several theses" which Hafenreffer and his colleagues excluded from the *Cosmographic Mystery*.

Between the Copernican astronomy and the literal interpretation of the Bible there was a head-on collision. The solution adopted by Hafenreffer and his colleagues sustained the authority of the Bible by suppressing the claim that it was reconcilable with Copernicanism, which was to be interpreted as a mathematical fiction. On the other hand, Kepler maintained the physical truth of the Copernican astronomy, without undermining the authority of the Bible, which was not to be interpreted literally.

Thus, when in the introduction to his *New Astronomy* Kepler discussed Joshua X. 12–14, he did not deny that a miracle occurred, since such a denial would have undermined the authority of the Bible. On the other hand, literal acceptance of the miracle as described ("the Sun stood still in the midst of heaven, and hasted not to go down about a whole day, and there was no day like that before it or after it") would have conflicted with Copernicanism, which regarded the Sun's immobility not as temporary or miraculous but as its permanent and normal condition. Kepler accepted the traditional belief that God created the Universe. But Kepler's God was a Copernican. When Joshua asked God to stop the Sun, "God readily understood from

---

[64] Kepler's *quomodo* was translated somewhat colloquially as *inwiefern* by Caspar–Dyck, I. 18. Had Kepler actually said that he showed "to what extent" the authority of the Bible remained, he would have implied that to some extent it did not. But his thesis was entirely different: the authority of the Bible, properly read, is unimpaired by Copernicanism.

[65] *GW*, XIII. 34: 43–45.

[66] In like manner Kepler inserted in the introduction of his *New Astronomy* (*GW*, III. 22: 27–24: 28) his student disputation attributing the Sun's apparent motion to the Earth. This is the disputation which was mentioned by Kepler in the second edition of his *Mysterium Cosmographicum* (Frankfurt am Main, 1621; *GW*, VIII. 28: 25–26).

Joshua's words what he wanted and complied by stopping the motion of the Earth, so that to Joshua the Sun seemed to stand still".[67] For the literal interpretation of the Bible Kepler substituted the principle that

> to teach mankind about ordinary things is not the purpose of Holy Writ, which speaks to people about these matters in a human way in order to be understood by them and uses popular concepts. . . . Why is it surprising, then, that Scripture also talks the language of human senses in situations where the reality of things differs from the perception?[68]

> Piety prevents many people from agreeing with Copernicus out of fear that the Holy Ghost speaking in Scripture will be branded as a liar if we say that the earth moves and the sun stands still. But these persons should bear in mind that we learn most things and the most important things with the sense of sight, and therefore cannot detach our speech from the visual sense. Thus, very many things happen every day when we talk the language of the sense of sight even though we know for a certainty that the situation is otherwise. . . . Thus, when we emerge from the narrow confines of a valley, we say that a wide field opens up before us.[69]

> "One generation passeth away, and another generation cometh, but the earth abideth forever" (says Ecclesiastes, [1:4]). Is this an argument with astronomers? [Kepler asks]. Are not people rather being reminded of their changeability by contrast with mankind's home, the earth, which always remains the same . . . ? You are not listening to any teaching in physics. This is a moral admonition concerning a matter which is self-evident and observed by everybody but taken into account too little. That is why it is emphasized.[70]

> Having discussed the authority of Holy Writ, I reply concisely to the beliefs of the [Church] Fathers regarding these aspects of Nature. In theology, the preponderance of authority must be weighed,[71] but in science the preponderance of reason. Therefore, revered is Lactantius, who denied that the earth is round. Revered is Augustine, who admitted [the earth's] sphericity but contested [the existence of] antipodes. Revered is the modern [Holy] Office, which concedes that the earth is small yet denies that it moves. On the other hand, the truth is revered more by me in proving scientifically, with all due respect to the Doctors of the Church, that the earth is spherical, inhabited all around by antipodes, most pitifully tiny and, finally, in motion among the celestial bodies.[72]

Kepler's clarion call, trumpeted to receptive ears, echoed and re-echoed down the corridors of the seventeenth century and thereafter. It demonstrated how unswerving allegiance to the scientific quest for truth could be combined in one and the same person with unwavering loyalty to religious tradition: accept the authority of the Bible in questions of morality, but do not regard it as the final word in science. Whereas Lutheran theologians in Kepler's time looked upon the Bible as a textbook

[67] *GW*, III. 30: 12–13; German translation in Max Caspar, *Johannes Kepler, Neue Astronomie* (Munich/Berlin, 1929), p. 30.

[68] *GW*, III. 29: 15–20; Caspar, *Neue Astronomie*, p. 29. Fifteen hundred years before Kepler, Rabbi Ishmael declared that the Bible speaks in the language of men. Although he was exclusively concerned with defending the redundancies in the Bible, Ishmael's dictum was later extrapolated beyond the domain of rhetoric to become a principle of exegesis: "We are . . . in duty bound to interpret the word of Scripture in a figurative sense . . . when the literal meaning contradicts a truth based on sense perception" (Henry Malter, *Saadia Gaon, His Life and Works*, Philadelphia, 1921, p. 234).

[69] *GW*, III. 28: 26–37; *Neue Astronomie*, pp. 28–29.

[70] *GW*, III. 31: 14–24; *Neue Astronomie*, p. 31.

[71] Reading *ponderanda* (*GW*, III. 33: 39).

[72] *GW*, III. 33: 37–34: 5; *Neue Astronomie*, p. 33.

of astronomy, subsequently their successors abandoned the literal interpretation of Scripture and moved in the direction indicated by Kepler's precept and example to that loftier plane on which the sincere practice of genuine religion could coexist with the acceptance of Copernicanism.

Meanwhile, on 5 March 1616 the Roman Catholic Church decreed the suspension of Copernicus' *Revolutions* "until corrected".[73] "Until explained" was the addition proposed by Kepler, who had taken that very task upon himself. In the second edition of his *Mysterium Cosmographicum* he inserted a footnote in which he said:

> In the introduction to my *Commentaries on the Motions of Mars* [as he often cited his *New Astronomy*] I have tried to show by arguments and examples how [Copernicus] is not in conflict with Scripture since his subject matter is utterly different. I have also expounded Copernicus' words more clearly at the end of Book I of my *Epitome of [Copernican] Astronomy*. I hope that in these passages I shall satisfy religious people, provided that they bring to the settlement of this question such an intellectual development and knowledge of astronomy that the glory of God's visible handiwork can be safely entrusted to them as guardians. Some weight is of course given to the word of God, but some is also given to the hand of God. Who would deny that God's word is attuned to its subject matter and for that reason to the popular speech of mankind? Hence, every deeply religious man will most carefully refrain from twisting God's word in the most obvious matters so that it denies God's handiwork in nature. Let anyone concerned about the praise of our Lord and Creator read Book V of my *Harmonics* [Linz, 1619]. When he has understood the most delicately harmonious coordination of the [celestial] motions, let him ask himself whether sufficiently correct and sufficiently productive reasons have been discovered for the agreement between the word of God and the hand of God, or whether there is any advantage in rejecting this agreement and by means of censorship destroying this glorification of the boundless beauty of the divine handiwork. The transmission of this glorification to the attention, however fleeting, of people without schooling, or even of the bulk of the educated, could never be accomplished by any decrees. The ignorant refuse to have respect for authority; they rush recklessly into a fight, relying on their numbers and the protection of tradition, which is impervious to the weapons of truth. But after the edge of the ax has been struck against iron, it does not cut wood any longer either. Let this be understood by anybody who is interested.[74]

"This impressive admonition is Kepler's last utterance regarding this matter", said Franz Hammer,[75] overlooking the later outburst in Kepler's *Somnium*, where he refers to an incident,[76]

> which happened to me, and yet not to me alone, but to a group of many like-minded persons. A theologian who professed the Augsburg Confession [the official creed of the Lutheran church] attacked us with great vehemence. He was convinced that he was drawing from Scripture weapons with which to assail us. Finally, becoming enraged by our self-defence, he

[73] The Latin text of the decree was republished by Giovanni Battista Riccioli, *Almagestum novum* (Bologna, 1651), I. 2, 496. The pope under whom Copernicus' *Revolutions* was suspended was Paul V, not Urban VIII (as Hammer mistakenly said in *GW*, VIII. 451).

[74] *GW*, VIII. 39: 26–40: 6: German translation by Caspar in *Weltgeheimnis*, pp. 38–39; of 39: 28–37, 40: 5–6, by Hammer in *GW*, VIII. 451.

[75] In *GW*, VIII. 451.

[76] Although Kepler dated this incident "in the year 1620", he wrote his account of it late in 1621 or in 1622, and therefore after the notes for his second edition of the *Mysterium Cosmographicum*; see Rosen, *Kepler's Somnium*, pp. xix–xx.

raised his voice and, calling on everything sacred as witness, he shouted that this teaching [Copernicanism] was "in conflict with all reason." Then I finally broke my stubborn silence (for up to that time I had been sitting there merely listening) and said: "No doubt it is this which impels even the ignorant men in your faction. For if the usefulness, necessity, and possibility of this teaching were understood by your narrow mind, you yourself would long ago have discounted the force of the arguments drawn from Scripture and sought out a suitable interpretation, as you are not infrequently in the habit of doing on other occasions. But now your mind is so feeble that you fail to see that on your side, too, there is some particle of reason. Therefore a teaching is not in conflict with all reason which is not in conflict with the reasoning of astronomers and physicists. For what someone fails to understand may be understood by someone else more familiar with the subject.[77]

It will be recalled that in the footnote cited above from the second edition of the *Mysterium Cosmographicum* Kepler referred to his *Epitome of Copernican Astronomy*, a title which he modeled after Mästlin's *Epitome of Astronomy*. But by inserting "Copernican", Kepler emphasized his open espousal of the new cosmology by contrast with his teacher who, as we saw above, believed it to be prudent pedagogical practice to keep Copernicanism out of an elementary textbook. In addition to the title, Kepler's *Epitome* followed Mästlin's in still another respect by being cast in the catechetical form of questions and answers. The last question in Book I asks:

What do you think is the proper reply to the religious and civil authorities of all ages and all ranks who regard as indisputable the anti-Copernican proposition that in the [apparent daily] first motion [of the heavens] the earth is stationary and the heavens move? Copernicus answers as follows:

1. Laymen control language and express what they see in familiar speech, [whereas] the philosopher seeks the truth which lies beneath the apparent forms of phenomena. Hence the philosopher's thinking not unreasonably differs from the judgment of laymen. 2. Some passages in Scripture are badly distorted for an astronomical purpose, and the opinions of those who habitually do so should be disregarded as utterly irrational. For with regard to astronomical topics which they had not learned in a systematic way, even the revered Church Fathers obviously spoke childishly sometimes and sought Scriptural authority for their mistake. An example is Lactantius, who could not believe that the Earth is round. The book of Job will appear to agree with Lactantius if it is sidetracked from the purpose of God's word to a philosophical fantasy.

This answer can be explained more fully. For astronomy discloses the causes of natural phenomena and takes within its purview the investigation of optical illusions. Much loftier subjects are treated by Holy Writ, which employs popular speech in order to be understood. Within this framework and with a different purpose in view, only in passing does Scripture touch on the appearances of natural phenomena as they are presented to [the sense of] sight, whence human speech originated, and proceed to do so even though it was perfectly clear to everybody that optical illusions were involved. Not even we astronomers cultivate astronomy with the intention of altering popular speech. Yet while it remains unchanged, we seek to open the doors of truth. That the planets are stationary or retrogress; the sun stands still, turns back, rises, sets, goes forth from one end of the heaven like a bridegroom coming out of his chamber and goes down into the other end [of the heaven], mounts to the midst of heaven, moves against certain valleys and mountains—these expressions are used by us along with laymen, that is, with the visual sense, even though not one of these locutions is literally true, as all astronomers agree. This is all the more reason not to require divinely inspired Scripture to

[77] Kepler, *Somnium*, p. 31; Rosen, *Kepler's Somnium*, pp. 36–38.

abandon the popular style of speech, weigh its words on the precision balance of natural science, confuse God's simple people with unfamiliar and inappropriate utterances about matters which are beyond the comprehension of those who are to be instructed, and thereby block their access to the far more elevated authentic goal of Scripture. Throughout the whole of this Book I [of my *Epitome of Copernican Astronomy*] may be seen traces of the Bible's popular expressions concerning the shape of the world and of the motions, about which there is no dispute. Why, then, are we in a sweat here only about the earth's motion?

Also dragged into this discussion are certain aspects which are not even open to inspection by the sense of sight but are entirely unrelated to our subject, as is proved by the character of the passage. They talk, for example, not about the earth's astronomical position or immobility but about its physical duration while in the meantime the animals on its surface perish and are born. Or they compare the stability of the land on which the animals move with their various motions. Or there is an allegory in which wars are settled and the public safety is signified by the establishment of the foundations of the earth.

3. As regards the authority of philosophers, Copernicus shows that at the very outset when astronomy was born there was no shortage of thinkers who declared the earth to be in motion from west to east, [namely,] Hicetas in Cicero, Philolaus and Ecphantus the Pythagoreans, and Heraclides of Pontus in Plutarch. To these may be added, from Archimedes and the afore-mentioned Plutarch, Aristarchus of Samos, a contemporary of Cleanthes, by whom he was accused of sacrilege before the court of the Areopagus because he had moved the altar of Vesta by asserting the motion of the Earth.

In our days all the most outstanding philosophers and astronomers agree with Copernicus. This sheet of ice is broken. Among the better [informed] voters we are winning. Almost the only obstacle for the others is irrational religion or fear of people like Cleanthes. But this is excessive. For even though none of our ancestors presented evidence for this truth, it should nevertheless have been accepted by the philosopher. For in Christian theology it is a topsy-turvy procedure first to seek a conclusion by reason and then afterwards weigh the authorities. Just so it is no less foolish in philosophy first to consider the authorities and then after-wards to pass on to reasoning.

Yet the authorities are cast aside by most educated people, whose knowledge is on a level not much higher than that of the uneducated. Acting by themselves and blinded by ignorance, first they condemn a discordant and unfamiliar doctrine as false. After deciding that it must be completely rejected and destroyed, then they look around for authorities, with whom they protect and arm themselves. On the other hand, they would make an exception of these same authorities, sacred and secular alike, in the same way as Copernicus, if they found them aligned on the side of the unconventional doctrine. They show this attitude in connection with the book of Job, Chapter 38, when anybody proves by means of it that the earth is flat, stretched to the tautness of a line, and resting on certain foundations, according to the literal meaning.[78]

The reference to Cleanthes' attack on Aristarchus should not be allowed to stand without comment, lest Kepler's version be accepted as historically correct. The highly compressed account presented here in the *Epitome* becomes more compre-hensible when it is placed in juxtaposition with Kepler's fourth objection to Aristotle. Kepler left in manuscript his own translation into German of Chapters 13–14, Book II, of Aristotle's *Heavens*, dealing with the Earth's shape, location, and motion or rest. To his translation Kepler added a number of objections, in the fourth of which he noted Aristotle's disparagement of the Pythagoreans, and continued:

[78] *GW*, VII. 99: 14–100: 31. In a letter to Mästlin on 8 December 1598 Kepler stated that "the book of Judges was written by various authors and is confused, as is obvious at its beginning and end" (*GW*, XIII. 258: 360–361).

Another mighty wiseacre[79] or philosopher, Cleanthes by name, was able to spend some
of his time plowing and digging, that is, making his living by working hard in the daytime
and drawing water, and the rest of his time studying. This [Cleanthes], I say, attended to poor
Aristarchus even more viciously [than Aristotle had to the Pythagoreans]. In the presence
of the pagan Athenian pope and priesthood Cleanthes made an accusation against Aristarchus,
charging him with a heresy punishable by death because he displaced the altar of the heathen
goddess Vesta. For Vesta was regarded as an earth goddess whose hearth, built in the middle
of every house, was sanctified and consecrated; whoever dishonored[80] the hearth committed
blasphemy. The hearth symbolized the Earth ("Herthum" in Old German).[81] Aristarchus
said that the Earth did not stand still like the hearth, nor was it in the middle of the universe
like the hearth in the middle of the house. Therefore he displaced the hearth of the goddess
and accordingly blasphemed her.[82]

Vesta was the Roman hearth-goddess, and thus the counterpart of Hestia, the
Greek goddess of the hearth. As the ancient sources of his knowledge about
Aristarchus, Kepler names Archimedes, who says nothing about Cleanthes, and Plutarch, who talked about the Cleanthes–Aristarchus situation in his *Face in the Moon*,
of which Kepler was then using a translation into Latin.[83] According to the *Face in
the Moon*, Cleanthes "thought" that an accusation should be brought against
Aristarchus, but Plutarch does not say, as Kepler does, that Cleanthes actually
brought such an accusation. Moreover, in Plutarch's account the accusation should
be brought by "the Greeks" in general, whereas Kepler places the accusation in
Athens, and specifically in the court of the Areopagus. This was the court before
which Cleanthes was brought to answer questions about how he made his living,
according to the biography of that leader of the Stoics by Diogenes Laertius.[84] He
was the source, direct or indirect, of Kepler's information that Cleanthes supported
himself by digging and drawing water. Hence, even though Diogenes Laertius is not
mentioned in our passage in the *Epitome* nor anywhere else in that work, Kepler
obtained some of his statements about Aristarchus from the *Lives of the Philosophers*.
In doing so, however, Kepler did not reread the relevant passage, but instead relied on
his memory, which in this case proved to be undependable. Thus, Plutarch's statement about Cleanthes' mere thought that Aristarchus should be accused of impiety by
the Greeks became intertwined, in Kepler's faulty recollection, with Diogenes

[79] Kepler's term *Witzkund* is registered neither in Jakob and Wilhelm Grimm, *Deutsches
Wörterbuch* (Leipzig, 1854–1960) nor in Hermann Fischer, *Schwäbisches Wörterbuch* (Tübingen,
1904–36).

[80] Kepler's word *entunehrete* is obsolete in his native Swabian dialect and not incorporated in
standard German.

[81] The supposed connection between *Erde* (earth) and *Herd* (hearth), as put forward here by
Kepler and elsewhere by others, is rejected by the Grimm *Wörterbuch*, III. 750.

[82] Frisch, VII. 744; Nikolaus Kopernikus, *Erster Entwurf seines Weltsystems sowie eine Ausein-
andersetzung Johannes Keplers mit Aristoteles*, ed. Fritz Rossmann (Munich, 1948; reprinted,
Darmstadt: Wissenschaftliche Buchgesellschaft, 1966), pp. 79–80.

[83] Wilhelm Xylander's Latin translation of Plutarch's *Face in the Moon*, as originally published
at Basel in 1570 and reprinted thereafter.

[84] *Lives of the Philosophers*, VII. 168–9.

Laertius' report that Cleanthes was actually haled before the Areopagus and that he wrote a (lost) work "Against Aristarchus". By this meandering path Kepler reached the unhistorical statement in his *Epitome* that Aristarchus was accused by Cleanthes of sacrilege before the court of the Areopagus.

Afterward, however, on account of his dissatisfaction with the existing Latin translation of Plutarch's *Face in the Moon*, Kepler decided to make his own, and for this purpose succeeded in obtaining a copy of the Greek text. When Kepler's Latin translation of Plutarch's *Face in the Moon* was published posthumously with his *Somnium* in 1634, his version correctly rendered Plutarch as saying "Cleanthes thought that the Greeks should bring an accusation of sacrilege against Aristarchus of Samos".[85] Here, as the attentive reader will have noticed, the actual accusation before the Areopagus of Athens has disappeared.

Just as in this relatively minor historical matter Kepler moved away from error toward truth, so in the vastly more important question of the relation between traditional religion and developing science he demonstrated that conflict was not inevitable. If the theologians would confine themselves to teaching morality and refrain from trying by dubious methods to extrapolate a constantly changing understanding of the physical universe from the Hebrew–Aramaic Bible and the Greek New Testament, treasured documents with no scientific pretensions, there could be genuine peace between the preacher in the pulpit and the thinker in the observatory. Today, as we look back over the tumultuous events since Kepler's birth, we should recognize that among his numerous contributions to the advancement of mankind we must accord a high place to his courageous and persistent struggle to convince his co-religionists, and all others who were willing to listen, that the Copernican astronomy was a giant stride forward in a direction that menaced no true believer.

NOTE A. Christian Frisch, ed., *Joannis Kepleri astronomi opera omnia* (Frankfurt am Main/ Erlangen, 1858–71), VIII, 651, citing I, 25, 37. At I, 25, Frisch began his reprint of Mästlin's introduction to Rheticus in Kepler's *Mysterium Cosmographicum*, and at I, 37, Frisch reproduced a part of this 30 October 1597 letter from Mästlin to Kepler. In neither document is the slightest evidence to be found of insincerity on the part of Mästlin. In fact, in 1606 "Mästlin became embroiled in a very serious and sharp quarrel with the [other] professors, who, on the basis of arguments drawn from Holy Writ, denied that the earth moves" (*GW*, XV. 355: 43–45). Frisch may have confused Mästlin's conduct with Hafenreffer's, as described by Kepler in the letter cited in note 45, above. Mästlin's conduct in 1606 disproves the assertion that "despite this intervention [by Mästlin on the Copernican side in Kepler's *Mysterium Cosmographicum* of 1596] thereafter he avoided any expression favorable to the new doctrine" (Ernst Zinner, *Geschichte und Bibliographie der astronomischen Literatur in Deutschland zur Zeit der Renaissance*, p. 36, Leipzig, 1941; 2nd ed., Stuttgart, 1964). More than a century earlier, Karl August Klüpfel had equally mistakenly said that although Mästlin was one of the earliest Copernicans "nevertheless he made the ancient system the basis of his writings for fear of offending the theologians" (*Geschichte und Beschreibung der Universität Tübingen*, p. 95, Tübingen, 1849).

[85] Kepler, *Somnium*, p. 106.

FIG.   4. Matthias Hafenreffer (1561–1619). (Landesbildstelle Württemberg, Stuttgart.)

Index